PLAYER 360

360

ELEVATE PLAYER ENGAGEMENT!

Sanjiv Goyal
Bryan Lindsey

ADROIT STUDIO

First Published in the United States in 2024 by Adroit Studio

Copyright © 2024–2025 by adroit

All rights reserved. This publication may not be reproduced, stored in any retrieval system, or transmitted in any form or by any means without prior written permission from the publisher. It may not be distributed in any other form of binding or cover other than the one in which it was published, nor may any condition be omitted for subsequent purchases.

ISBN: 979-8343351972
Publisher's Website: www.adroit.cc

DEDICATION

In heartfelt dedication to our team and cherished customers, whose unwavering support has been the cornerstone of our success over the decades.

CONTENTS

Preface 7

Introduction 10

The Disconnect 15

Engagement Reimagined 23

More Than One Channel 30

The Value Equation 40

In the Moment 52

Loyalty Unlocked 66

AI at Play 81

The Future Arrives 90

Data That Matters 99

Data's Dilemma 108

Privacy at Stake 117

Responsible Gaming 127

Effortless Operations 137

The Human Factor 147

Gamification's Edge 156

Adapting to Win 170

Beyond the Games 179

Learning from Others 188

Marketing Reinvented 197

The Next Frontier 206

The Call 215

Appendix 224

ELEVATE PLAYER ENGAGEMENT

PREFACE

The casino industry is a fascinating world filled with energy, excitement, and the constant challenge of keeping players engaged. Over the years, the way casinos interact with their players has evolved significantly. What was once a business driven by gut instinct and personal relationships has transformed into a highly sophisticated operation where data, technology, and strategy play crucial roles. In this book, we explore one of the most groundbreaking changes in this industry—how Artificial Intelligence (AI) is revolutionizing player development.

As someone deeply involved in the casino industry, you've likely seen firsthand how technology has changed how we do business. Digital platforms and data analytics have opened up new possibilities, but they have also brought new challenges. The key to staying ahead in this competitive environment lies in effectively leveraging these tools. AI is at the forefront of this technological revolution, offering unimaginable insights and capabilities just a few years ago.

Player development has always been about understanding your players, anticipating their needs, and offering them experiences that keep them coming back. Traditionally, this was achieved through personal relationships, with hosts and player development executives relying on their knowledge of individual players to build loyalty. While these personal touches are still important, the scale and complexity of today's casino operations require more than just intuition. This is where AI steps in, offering a level of precision and efficiency that is transforming the industry.

AI enables casinos to analyze vast amounts of data in real-time, providing insights that can help you understand your players on a deeper level. It can predict player behavior, identify trends, and even suggest personalized offers that are tailored to each player's preferences. This level of personalization was once only possible in a one-on-one setting, but AI allows you to scale it across your entire player base. This means you can deliver the right message, at the right time, to the right player, every time.

But AI is not just about data and analytics. It's also about enhancing the human element of player development. By automating routine tasks, AI frees up your team to focus on what they do best—building relationships with your players. It provides them with the tools and information they need to make smarter decisions, faster. This combination of human intuition and machine precision creates a powerful synergy that can drive your player development efforts to new heights.

This book is designed to be a practical guide for casino executives, player development teams, and anyone interested in the future of the casino industry. It explores how AI can be integrated into player development strategies, from predictive analytics to real-time engagement systems. We'll look at the challenges and opportunities of AI adoption and provide actionable insights you can apply to your operations.

As you read through these chapters, you'll discover how AI can help you retain your most valuable players and attract new ones. You'll learn how to create more personalized, engaging experiences that keep your players returning for more. And most importantly, you'll see how AI can give you a competitive edge in a constantly evolving industry.

The journey to fully integrating AI into your player development strategies may seem daunting, but the rewards are well worth the effort. Whether you're just starting out or looking to refine your existing processes, this book will provide you with the knowledge and tools you need to succeed. The future of player development is here, and it's powered by AI.

INTRODUCTION

The casino industry is standing on the precipice of a profound transformation. The glitzy days when casinos could rely on lavish promotions, thrilling games, and extravagant entertainment to secure success are quickly fading. Today's players demand more. They want personalized experiences that align with their unique preferences, seamless interactions that follow them wherever they go, and rewards that make them feel genuinely valued—not just as players but as individuals. The competitive edge no longer lies in who has the biggest jackpots or the flashiest show, but in who can deliver an experience that feels personal, dynamic, and immersive.

As the landscape shifts, the challenge for casinos isn't merely offering more games or grander attractions but **redefining how they engage with players**. The journey a player takes today is far more complex than a decade ago. A slot machine and a table game are no longer just singular experiences; they are part of a broader, continuous journey. It might begin on a player's mobile device while at home, continue an online gaming platform during a commute, and culminate at the casino bar after a day at the tables. Players expect every interaction to feel connected, fluid, and personalized throughout this journey.

Casino operators must ask how they can stay ahead in this rapidly evolving landscape. The answer lies not just in technology but in adopting a new **philosophy**, a mindset that focuses on the player's entire journey—**the Player 360 mindset**. This philosophy embraces **integration**, personalization, and **real-time engagement**, transforming the way casinos think about their players and ensuring that every step of their journey feels connected.

A Holistic Approach

The casino of the future will no longer be a single-dimensional space where guests come simply to gamble. It will evolve into a **multifaceted environment**, blending physical and digital worlds to offer a seamless array of experiences that transcend the gaming floor. From the moment a player interacts with a casino's mobile app to the time they step onto the property, and later when they engage with online gaming platforms, each touchpoint becomes a critical moment to **understand** and **deepen engagement**.

However, building this future requires a **holistic approach to casino transformation**. Imagine a player opening a mobile app to check their loyalty rewards, browse offers, or play a few rounds of digital slots. As they arrive at the physical casino, everything they do online follows them: their preferences for favorite games, past wagers, and even their favorite drink orders are known and ready. They feel **recognized**, **valued**, and **catered to** in a deeply personal yet unobtrusive way.

This level of engagement demands more than traditional systems can offer. It requires a **360-degree view of the player**—an approach that integrates every interaction across every platform. The key to success is **transforming data into actionable insights** that guide real-time decisions. For instance, if a high-value player frequently checks their loyalty status via a mobile app but hasn't visited the casino in several weeks, the system can trigger a **personalized offer**—a free hotel stay, exclusive dinner reservation, or tickets to a special event—designed to encourage their return.

This isn't about **reactive engagement**; it's about being **proactive**. It's about knowing where a player is in their journey and offering them an experience that resonates in the moment.

A 360-Degree Experience

At the heart of the **Player 360 mindset** is the seamless integration of player behavior across multiple platforms and environments. In the past, player's interactions were limited to what they did within the physical walls of the casino. But in today's interconnected world, **player behavior spans far beyond the property**, creating valuable insights into what players enjoy and how best to engage them.

Players no longer interact with casinos in isolation; they engage across multiple touchpoints—**mobile apps**, **online gaming platforms**, and **on-property experiences**. This holistic view of the player journey allows casinos to **anticipate needs**, offer **personalized experiences**, and foster **loyalty**. Consider a player who enjoys online poker and regularly visits the casino's upscale dining options. By integrating data across both digital and in-person interactions, casinos can offer this player an **exclusive poker tournament invitation** and a dinner reservation at their favorite restaurant, creating a seamless and meaningful experience.

The **potential** of this approach is vast, offering casinos the ability to not only collect data but also to **use it to drive engagement** in ways that feel personal and compelling. Every touchpoint becomes an opportunity to build a more robust relationship with the player, and every interaction becomes part of the larger journey.

From Data to Action

The future of casino operations lies in **data-driven personalization**. Casinos already collect vast amounts of data, from gaming habits and loyalty program activity to dining preferences and mobile app usage. But this data is only valuable if it is **used effectively**—not just to understand players but to **anticipate their needs**.

Consider the player who loves high-stakes poker and frequently books suites at the hotel. This data enables the casino to offer **personalized promotions**, such as access to exclusive poker tournaments or luxury spa experiences. By aligning offers with the player's known preferences, the casino creates a sense of **value** and **recognition**, strengthening the relationship and ensuring future engagement.

This level of **personalization** goes far beyond the "one-size-fits-all" promotions of the past. The modern player expects **relevance**—whether it's a free play bonus for an online game or a dinner reservation based on past visits, the offer must resonate with their interests. Personalized engagement isn't just about data; it's about building **long-term relationships** that deepen loyalty and enhance the overall player experience.

An Integrated Ecosystem

The casino industry is rapidly evolving, and the future lies in creating an **integrated ecosystem** that bridges online and offline experiences. The days of disconnected platforms are over; players expect **seamless engagement** across every touchpoint, whether they're using a mobile app, engaging in online gaming, or stepping onto the casino floor.

To thrive in this new era, casinos must adopt a **holistic mindset** that places the player at the center of every decision. It's not just about technology—it's about understanding that **every interaction matters** and that **every player touchpoint is an opportunity** to deliver value.

This is more than a strategy—it's a **philosophy**. It's about building an ecosystem that fosters **loyalty**, anticipates **needs**, and provides players with the **personalized experiences** they crave. It's about offering more than just games; it's about creating a **vision** for the future where every interaction is part of a connected, immersive journey.

As you dive deeper into this book, you'll explore the principles, technologies, and philosophies that will guide casinos **beyond 2030**. This is a roadmap for becoming an **outlier**—a leader in the new era of gaming where **personalization**, **data integration**, and **player-centric engagement** define the future. The transformation is already happening, and those who embrace this vision will be the ones to lead the charge.

Are you ready to build tomorrow's casino today? The **Player 360 mindset** is more than a concept—it's the blueprint for success in the next generation of gaming. It's time to redefine what's possible and to think beyond the games themselves.

The future of the casino industry is here. Will you answer the call?

1.

THE DISCONNECT

The problem of fragmented player experiences

Emma stepped into the grand casino lobby with a thrill of excitement. The vibrant lights, the rhythmic hum of slot machines, and the soft sound of chips stacking at the tables filled the air. But as Emma moved through the floor, trying her hand at a few games, she couldn't help but feel distant from the experience. She had visited this casino many times before, yet nothing about her visit felt personal. The emails she had received promoting offers seemed disconnected from her preferences. The mobile app didn't reflect her recent activity. Despite her frequent visits, none of the staff recognized her or acknowledged her loyalty.

Emma's experience is not uncommon. It's a symptom of a larger problem that casinos around the world face today—a growing disconnect between players and the places that should feel like their second home. While casinos are designed to deliver excitement, entertainment, and an immersive experience, they are often missing the one thing that truly makes players feel valued: personalization.

The Fragmented Player Experience

Today's casinos are not just confined to the physical space of the gaming floor. The modern player engages with casinos through a multitude of channels—mobile apps, websites, social media, email marketing, and of course, the physical venue. Each of these touchpoints offers an opportunity to connect with players, but too often these interactions exist in silos.

For Emma, her experience was fragmented. She had received promotional emails, but they offered her discounts on games she rarely played. The app she used to check her rewards didn't sync with her recent activity, leaving her unsure of what benefits she had accrued. When she arrived at the casino, no one seemed to recognize her as a loyal customer. Every interaction felt disconnected, as though the casino was unaware of her past visits, her preferences, and the loyalty she had shown.

This fragmentation of experiences is more than an inconvenience—it undermines the essence of what keeps players coming back. Casinos should be places where players feel known, valued, and appreciated. But when the touchpoints don't align, players like Emma are left feeling like just another face in the crowd. The result? Players become disengaged, less loyal, and more likely to seek entertainment elsewhere.

A New Era of Expectations

In today's digital age, players expect more than ever before. They've grown accustomed to a world where personalization is the norm. Consider services like Netflix, Amazon, and Spotify—each of these platforms uses data to tailor recommendations based on individual preferences, viewing habits, or purchase history. They create experiences that feel tailored, relevant, and timely.

Casinos, however, have often lagged behind this trend. While the gaming floor may have evolved with new slot machines and technologies, the overall player experience has not kept pace with the personalization players expect in their everyday lives. Instead of receiving offers that align with their gaming preferences, players often receive generic promotions, making them feel as though they are not understood or valued as individuals.

Take Emma, for example. She prefers slot machines and occasionally dabbles in blackjack. Yet the emails she receives from the casino frequently promote poker tournaments or other games that don't align with her interests. This kind of mismatch between player preferences and promotional efforts signals to Emma that the casino isn't paying attention. Over time, this erodes her loyalty, making her less likely to engage with future offers and more likely to seek a more personalized experience elsewhere.

The Financial Impact of Disconnect

The consequences of this disconnect are not just about player satisfaction—they have real financial implications for casinos. Casinos invest millions in marketing campaigns, email promotions, loyalty programs, and player engagement initiatives. But when these efforts are not tailored to individual players, much of that investment is wasted.

Consider what happens when Emma receives an email offering free slot play. However, the offer is not available on her favorite machines. Instead of being excited about the promotion, Emma feels frustrated. She was eager to engage, but the offer doesn't align with her preferences. In this case, the casino not only loses out on Emma's engagement with the promotion, but it also risks damaging her perception of the casino as a place that understands her needs.

Over time, the cumulative effect of these missed opportunities can be profound. Players who feel disconnected are less likely to engage with promotions, less likely to return to the casino, and less likely to spend as much money during their visits. Studies have shown that personalized marketing can significantly increase player engagement and loyalty, leading to longer visits, more frequent play, and increased spend per visit.

But the opposite is also true. When casinos fail to personalize the experience, they risk losing not just individual players like Emma, but entire segments of their customer base. And as competition in the gaming industry intensifies—both from other casinos and from alternative forms of entertainment—casinos cannot afford to overlook the importance of connection and personalization.

Why the Disconnect Happens

One of the main reasons why this disconnect occurs is because many casinos operate with fragmented, siloed systems. Each department—marketing, loyalty programs, gaming operations—often uses its own set of tools to manage player data, promotions, and engagement. These systems don't always communicate with each other effectively, which prevents the casino from creating a unified view of each player.

In Emma's case, the marketing team may have her email address and her preferences for certain types of promotions, but they don't have access to real-time data about her behavior on the gaming floor. So, they send her offers that don't match her current interests. Similarly, the mobile app she uses may not be integrated with the loyalty system, meaning that she can't see real-time updates on her rewards. And when she arrives at the casino, the staff on the gaming floor doesn't have access to her past visit history, so they treat her like a new player rather than a loyal guest.

The problem is that while casinos have vast amounts of data on their players, they often don't have the infrastructure or systems in place to effectively use that data to enhance the player experience. This is a missed opportunity. By integrating their systems and creating a holistic view of each player, casinos can offer personalized, relevant experiences that resonate with players on a deeper level.

The Importance of the Human Element

While technology plays a crucial role in addressing the disconnect, it's important to remember that the human element is still essential in creating a meaningful player experience. Casinos are not just about machines and algorithms—they are about people. Hosts, dealers, and staff are often the face of the casino, and their ability to engage with players can make a huge difference in how players feel about their experience.

For example, a good casino host can transform a player's visit by offering personalized attention, making suggestions based on the player's preferences, and ensuring that the player feels valued. But without real-time data about a player's history and preferences, even the best hosts are limited in what they can offer.

In Emma's case, her host didn't know that she preferred slot machines over table games. They didn't know that she had visited the casino three times in the past month or that she had recently participated in a promotion. Without this information, the host couldn't provide the level of personalized service that might have deepened Emma's loyalty to the casino.

This highlights an important point: while technology is critical in solving the disconnect, it should be used to empower human interactions, not replace them. By giving hosts and staff the tools and information they need, casinos can create a more personal, connected experience for their players.

The Urgency of Change

The gaming industry is at a critical juncture. Competitors are not just other casinos, but also other forms of entertainment—from mobile gaming apps to streaming services to social media platforms. To remain relevant and competitive, casinos must address the disconnect head-on.

This means rethinking how they approach player engagement, moving away from siloed systems and one-size-fits-all marketing campaigns, and embracing a more integrated, personalized approach. It requires a shift in mindset—from seeing players as anonymous customers to viewing them as individuals with unique preferences, habits, and interests.

Casinos that can make this shift will not only enhance player satisfaction but also see tangible improvements in player retention, engagement, and revenue. Players like Emma, who feel recognized and valued, will return more often, spend more during each visit, and become advocates for the casino, encouraging their friends and family to visit as well.

A Vision for Reconnection

This book will explore how casinos can bridge the gap and reconnect with their players. We'll dive into strategies for creating seamless, personalized experiences across all touchpoints—whether through targeted marketing campaigns, real-time engagement on the gaming floor, or innovative loyalty programs.

We'll also explore the role of emerging technologies—such as artificial intelligence, machine learning, and predictive analytics—in transforming the player experience. These technologies offer new opportunities to anticipate player needs, deliver personalized offers, and create moments of delight that keep players coming back.

At the same time, we'll address the critical issues of data privacy and security. As casinos collect more data on their players, they must also ensure that they are using that data ethically and responsibly. Players need to feel confident that their information is being protected, and that their privacy is being respected.

The disconnect is real, but it's also solvable. By embracing a holistic, player-centric approach, casinos can create the kind of connected, personalized experiences that will not only keep players like Emma engaged but also drive long-term success.

Key Takeaways

1. **Fragmented experiences drive player disengagement**: When touchpoints across a player's journey aren't integrated, it creates a disjointed experience that leaves players feeling undervalued and disconnected.
2. **Players expect personalization**: Today's consumers are used to personalized interactions across industries. Casinos that fail to offer relevant, customized experiences risk losing players to competitors.
3. **The cost of disconnection is real**: Generic marketing and impersonal engagement lead to wasted opportunities, reduced player loyalty, and lower revenue. Personalization can significantly increase player engagement and spending.
4. **Siloed systems are the main cause**: Many casinos operate with fragmented systems that don't communicate with each other. This prevents the integration of player data, limiting the ability to offer personalized experiences.
5. **The human touch is critical**: While technology is important, human interactions remain essential. Empowering staff with real-time data and insights allows them to provide the

ELEVATE PLAYER ENGAGEMENT

personalized service that keeps players coming back.

2.

ENGAGEMENT REIMAGINED

Campaigns do reshape player engagement!

Emma's frustration wasn't just about the disconnect between what she wanted and what the casino offered—it was deeper. As she moved from the slots to the tables, she couldn't shake the feeling that her time at the casino was like everyone else's. The promotions were the same; the events felt generic, and nothing was tailored specifically to her. The casino seemed to focus on blasting out mass offers, but none of them spoke to her directly. Emma felt part of an outdated system, a machine that churned out loyalty points but didn't care to understand her preferences or habits. This is the reality for many players today. The casino experience hasn't evolved in ways that meet the new expectations of the modern player.

The Old Approach to Engagement

For years, casinos have relied on what could be called a "blanket approach" to player engagement. This method focuses on casting a wide net, offering mass promotions, and hoping that a decent percentage of players will engage. Traditional engagement strategies—loyalty programs, point systems, and periodic offers—have been the backbone of casino marketing. However, as player expectations have evolved, these one-size-fits-all approaches have shown their limitations.

In this old model, players like Emma would receive the same email promotions, attend the same events, and participate in the same loyalty program as everyone else. The logic behind this is simple: if a casino can offer enough players the same promotion, some of them are bound to take the bait. However, this method is inherently flawed because it treats players as a homogenous group rather than as individuals with unique preferences, habits, and behaviors.

While the blanket approach worked well in the past when players had fewer choices for entertainment, today's landscape is different. Players are no longer confined to traditional gaming floors. They can access online casinos, mobile apps, and countless other entertainment options at their fingertips. To stand out, casinos must move beyond this old model of engagement.

A New Vision for Player Engagement

To reimagine engagement, casinos must begin by acknowledging the diversity of their player base. No two players are exactly alike. Some prefer high-stakes table games, while others like casual slot play. Some visit the casino for the nightlife, while others come purely for gaming. Players also have varying levels of experience—from novice to seasoned gambler—and their motivations for visiting the casino may differ dramatically. Some are there to win big; others are there for the social experience or relaxation.

The key to effective player engagement is recognizing these differences and crafting a strategy that appeals to players on an individual level. This new approach involves not only understanding player preferences but also predicting their needs and tailoring offers to their unique habits. This shift from mass marketing to personalized engagement represents a fundamental rethinking of how casinos interact with their customers.

For Emma, this means receiving offers that reflect her actual interests. Instead of a generic promotion for a poker tournament, she might receive a personalized offer for a free spin on her favorite slot machine. When she walks onto the gaming floor, her app could update with a notification welcoming her back and suggesting a new slot game that aligns with her previous play history. These small touches make Emma feel valued, understood, and more likely to engage with the casino in meaningful ways.

The Role of Data in Personalization

Personalized engagement starts with data. Casinos are uniquely positioned to collect vast amounts of information on their players—from their gaming habits and spending patterns to the promotions they engage with and the games they prefer. However, collecting data is only the first step. The real value lies in how casinos use this data to deliver personalized experiences.

In the past, many casinos have struggled to make the most of their data. Information about players is often siloed across different departments, making it difficult to get a complete view of an individual player's preferences and behaviors. Marketing teams might have access to email data, but not to real-time gaming floor activity. The loyalty program might track points, but it doesn't always capture a player's engagement with digital platforms like mobile apps or websites.

To reimagine engagement, casinos must break down these silos and integrate their data into a centralized system. By doing so, they can create a unified profile for each player, allowing them to tailor offers and experiences based on a holistic understanding of that player's preferences. This means not only recognizing which games a player likes to play, but also understanding when and how they prefer to engage with the casino.

For example, Emma might be a regular visitor on Friday nights, spending most of her time at the slot machines. With integrated data, the casino can send her a targeted offer for a free drink or bonus play, timed to arrive just before she visits. If she recently tried out a new game on the app, the casino could follow up with a personalized recommendation for a similar game when she arrives in person. This kind of dynamic, real-time engagement transforms the player experience from generic to bespoke.

Creating Meaningful Moments

Reimagining engagement isn't just about sending better promotions—it's about creating meaningful moments for players that deepen their connection to the casino. These moments might take the form of personalized offers, but they can also be subtle touches that make the player feel known and valued. It could be something as simple as having a staff member greet a player by name or offering a special birthday reward that acknowledges the player's loyalty.

For Emma, meaningful moments might include a personalized welcome message when she walks onto the gaming floor, tailored recommendations for games she hasn't tried but would likely enjoy, or even a surprise bonus for reaching a certain milestone in her loyalty points. These interactions make her feel like more than just a number—

they make her feel like the casino knows and appreciates her.

By focusing on creating meaningful moments, casinos can build deeper relationships with their players. Players who feel valued are more likely to return, more likely to engage with promotions, and more likely to become advocates for the casino. These players are also more likely to stay loyal, even in the face of competing entertainment options.

The Evolution of Loyalty Programs

At the heart of player engagement in casinos has traditionally been the loyalty program. These programs are designed to reward frequent visitors with points, prizes, and special offers. But in many cases, loyalty programs have become static and transactional. Players earn points, redeem rewards, and move on. There's little emotional connection or personalized interaction.

To reimagine engagement, casinos must evolve their loyalty programs beyond simple points and rewards. A truly effective loyalty program doesn't just incentivize play—it creates a deeper emotional connection with the player by offering experiences that are tailored to their preferences and behaviors.

Consider how airlines have evolved their frequent flyer programs. While points and miles are still a major component, airlines now focus on creating tiered experiences offering personalized perks based on travelers' preferences. Casinos can take a similar approach, offering personalized rewards based not just on points accrued but on the individual's specific gaming habits, frequency of visits, and level of engagement.

For Emma, this might mean that as she plays her favorite slot machines, she earns rewards specifically designed for her—free play on certain games she enjoys or access to exclusive events that match her interests. The loyalty program would go beyond mere points, offering a curated experience that makes her feel special and appreciated.

Moving Beyond Transactional Relationships

The shift from transactional relationships to emotional connections is at the core of reimagining engagement. In the past, casinos have focused on incentivizing transactions—spend more, get more points. But today's players, like Emma, want more than just rewards for their spend. They want to feel connected to the casino in a way that goes beyond the transactional.

To achieve this, casinos must focus on creating emotional connections with their players. This means understanding not only what games they like to play, but also why they play. Are they looking for a thrill? A social experience? A relaxing escape? By tapping into the emotional drivers behind player behavior, casinos can tailor their engagement strategies to resonate more deeply.

For Emma, it's not just about the free spins or bonus points—it's about the experience. She comes to the casino to unwind, to enjoy herself, and to have fun. By understanding her motivations, the casino can create experiences that make her feel more connected, engaged, and loyal.

Key Takeaways

1. **The blanket approach is outdated**: Traditional mass marketing and loyalty programs are no longer effective in

meeting the needs of today's diverse player base. Casinos must shift from one-size-fits-all promotions to personalized engagement strategies.

2. **Data is the key to personalization**: To deliver personalized experiences casinos must integrate their data across all touchpoints and create a unified profile for each player. This allows for more targeted, relevant offers and a seamless experience.

3. **Creating meaningful moments matters**: Engagement isn't just about sending promotions—it's about creating moments that make players feel known and valued. Personal touches, tailored offers, and real-time engagement build emotional connections with players.

4. **Loyalty programs must evolve.** They must move beyond points and rewards to offer personalized and emotionally engaging experiences. This will strengthen connections with players and encourage long-term loyalty.

5. **Emotional connections trump transactions**: Players want more than just rewards for their spending—they want experiences that resonate with their motivations and desires. Casinos that focus on creating emotional connections will see greater loyalty and engagement.

3.

MORE THAN ONE CHANNEL

Multi-Channel Strategies for Deeper Engagement

Emma sits at home, excited to play her favorite online slot game after a fun night at the casino a few days ago. She logs into the casino's app, checks her loyalty points, and notices that her progress from the casino floor has been seamlessly integrated into the app. There's a special offer just for her—a bonus for using the app and visiting the casino in person. Emma smiles, feeling that the casino understands her preferences and rewards her for engaging on multiple fronts. Whether she's playing at the casino, on her phone, or her laptop, it feels like one continuous, connected experience. The casino has mastered the art of **multi-channel engagement***, and Emma is hooked.*

The Importance

In today's digital world, players no longer engage with casinos through a single channel. Players like Emma move between physical locations, mobile apps, websites, and social media, often expecting the same high-quality experience at each touchpoint. This shift in player behavior has made it essential for casinos to adopt **multi-channel strategies**—approaches that engage players seamlessly across multiple platforms, providing a unified and consistent experience that encourages deeper engagement.

A multi-channel strategy involves connecting all the ways players can interact with the casino—whether that's through the physical casino, a mobile app, a website, or even social media. Each channel offers unique opportunities to engage players, but it's the ability to **link these channels** that creates a truly engaging experience. Players no longer see the casino as just a building—they see it as an ecosystem of interactions that fit their lifestyle, whether they're at home, on the move, or on the gaming floor.

This chapter explores the power of multi-channel strategies in the gaming industry, focusing on how casinos can deepen player engagement by connecting physical, digital, and mobile experiences. We'll also look at how casinos can use data, personalization, and real-time interactions to create a cohesive experience that rewards players like Emma across all channels.

What is it?

A **multi-channel strategy** in the casino industry refers to the practice of offering players multiple avenues to engage with the casino, while ensuring that each channel is interconnected and provides a consistent experience. Rather than treating the physical casino, mobile app, and online platform as separate entities, a multi-channel approach ensures that Emma's journey is integrated. Whether she's playing on the casino floor, using her mobile phone, or interacting with the casino through social media, her experience is seamless and cohesive.

For Emma, a multi-channel strategy means she can easily switch between channels without losing progress or access to promotions. It's about offering flexibility while creating a unified player experience. For casinos, multi-channel strategies provide more touchpoints to engage players, which increases the potential for higher engagement, loyalty, and revenue.

Benefits

1. Increased Engagement Across Platforms: By offering players the ability to engage across multiple channels, casinos can significantly increase engagement. Emma, for instance, might visit the physical casino once a month but engage with the mobile app or online platform multiple times per week. This gives the casino more opportunities to interact with her, keeping her connected to the brand even when she's not physically present.

For Emma, the multi-channel strategy offers flexibility. If she's busy during the day but wants to play a few rounds on her favorite slot machine, she can use the casino's mobile app. If she's out with friends, she can use the app to check her loyalty points or see what promotions are available when she returns to the casino. By staying connected across multiple platforms, Emma remains engaged with the casino throughout the week.

2. Seamless Cross-Channel Experience: One of the key advantages of a multi-channel strategy is the ability to offer a seamless experience across all platforms. For Emma, this means that her progress in games, loyalty points, and bonuses are synchronized no matter where she plays. If she earns points at the physical casino, those points are immediately reflected in the mobile app, allowing her to redeem rewards or continue earning bonuses when she switches to online play.

This seamless experience makes Emma feel valued and encourages her to use multiple channels. She knows that she won't miss out on any rewards or progress, no matter where or when she chooses to engage with the casino.

3. Personalization Across Channels: A well-executed multi-channel strategy allows casinos to personalize the player experience across platforms. With data collected from Emma's activities on different channels, the casino can tailor promotions, game recommendations, and rewards to match her preferences, making her feel understood and appreciated.

For example, if Emma tends to play table games when she visits the physical casino but enjoys slot machines when she's using the app, the casino can use this data to send her personalized offers for both experiences. This keeps her engaged across channels and ensures that she always feels like the casino is offering her something relevant to her interests.

4. Greater Flexibility and Convenience for Players: Multi-channel strategies give players the freedom to engage with the casino whenever and however they want. Whether Emma prefers playing games in the casino, on her phone during her commute, or at home on her laptop, the casino offers a flexible experience that fits her lifestyle. This convenience is a major driver of loyalty and engagement, as it allows players to interact with the casino on their terms.

For Emma, this flexibility means she can choose when and where she plays, without feeling restricted to a single platform. If she's traveling, she can still engage with her favorite games through the mobile app, and when she returns to the casino, all her progress will be waiting for her.

5. Strengthened Loyalty and Retention: A multi-channel strategy helps strengthen player loyalty by providing consistent rewards and recognition across platforms. By ensuring that Emma's loyalty points and rewards are easily accessible on every platform, the casino reinforces her connection to the brand. Emma is more likely to stay loyal to the casino because she feels that her engagement is recognized and rewarded, no matter how she chooses to interact with the casino.

Additionally, offering promotions that span multiple channels—such as earning extra points for playing both online and in person—encourages Emma to use multiple platforms, increasing her overall engagement and loyalty

Key Multi-Channel Strategies

To fully leverage the power of multi-channel engagement, casinos must implement several key strategies that connect the physical, digital, and mobile experiences in meaningful ways. These strategies ensure that players like Emma receive a consistent, personalized experience across all touchpoints, keeping them engaged and loyal to the brand.

1. Integrated Loyalty Programs: A successful multi-channel strategy relies on an integrated loyalty program that tracks and rewards player activity across all channels. For Emma, this means that her loyalty points, rewards, and status levels are updated in real time, whether she's playing on the casino floor, using the app, or engaging with online games. The loyalty program must be accessible across platforms, allowing Emma to view her progress, redeem rewards, and participate in promotions from any device.

An integrated loyalty program not only incentivizes Emma to engage across multiple channels but also makes her feel valued. She knows that her efforts are recognized, and she is rewarded no matter where or how she plays.

2. Cross-Channel Promotions: One of the most effective ways to drive engagement across multiple channels is through cross-channel promotions. For example, the casino could offer Emma a promotion that encourages her to play both online and in person, such as earning bonus points for playing a certain number of games on the app and then visiting the physical casino to redeem a reward. These promotions incentivize Emma to use multiple platforms, increasing her overall engagement with the casino.

Cross-channel promotions also create a sense of continuity in Emma's experience, making her feel that her actions on one platform are connected to her progress on another. This seamless transition encourages her to explore all of the casino's offerings, rather than limiting herself to a single channel.

3. Personalized Content and Offers: Personalization is a critical element of any multi-channel strategy. By analyzing Emma's activity across platforms, the casino can deliver personalized content, offers, and recommendations that align with her preferences. For instance, if Emma plays specific games more frequently on her mobile app, the casino can send her tailored promotions for those games, both in the app and in the physical casino.

Personalized offers not only increase Emma's engagement but also enhance her sense of loyalty. She feels that the casino understands her preferences and is offering her rewards that are relevant to her gaming habits. This personal touch keeps her coming back for more.

4. Consistent Messaging and Branding: To create a cohesive player journey, it's essential that the messaging and branding remain consistent across all channels. Emma should experience the same level of service, tone, and design whether she's using the mobile app, playing on the website, or visiting the physical casino. Consistent branding reinforces the casino's identity and ensures that Emma recognizes the casino's presence on every platform.

For example, the casino's mobile app should reflect the same look and feel as its physical environment, creating a unified experience that strengthens brand recognition and loyalty. Messaging, too, should be consistent, with promotions and offers delivered in a unified voice across channels.

5. Real-Time Engagement: Real-time engagement is key to keeping players like Emma connected across multiple channels. The casino should use push notifications, emails, and in-app messages to deliver timely promotions, reminders, and updates based on Emma's activity. For instance, if Emma hasn't visited the physical casino in a while but has been actively playing on the app, the casino could send her a personalized message offering a reward for her next in-person visit.

Real-time engagement ensures that Emma feels connected to the casino at all times, even when she's not physically present. It also helps drive immediate action, encouraging Emma to take advantage of promotions or rewards while they are still relevant.

Challenges in Multi-Channel Strategies

While multi-channel strategies offer numerous benefits, implementing them can also present challenges. Casinos must navigate technological complexities, data integration issues, and the need for consistent branding to ensure a seamless experience.

1. Data Integration:
One of the biggest challenges in executing a multi-channel strategy is integrating data from multiple platforms. Casinos

need to ensure that player data, such as gameplay history, loyalty points, and rewards, are synchronized across all channels in real time. This requires advanced data management systems, such as customer data platforms (CDPs), that can centralize information and provide a unified view of each player's activity.

2. Maintaining Consistency Across Channels:
Creating a consistent experience across platforms requires meticulous attention to branding, messaging, and service. Casinos must ensure that players like Emma receive the same quality of service, whether they're interacting with the mobile app, website, or casino staff. This consistency is essential for building trust and loyalty.

3. Personalization at Scale:
While personalization is key to multi-channel success, delivering personalized experiences at scale can be challenging. Casinos must use AI and data analytics to analyze player behavior across platforms and deliver personalized offers that are relevant and timely. This requires sophisticated technology capable of handling large volumes of data and interactions.

The Future of Multi-Channel Strategies in Casinos

As technology continues to evolve, the future of multi-channel strategies will become even more dynamic and interconnected. Emerging technologies like artificial intelligence, machine learning, and cloud computing will allow casinos to create more personalized, real-time experiences that engage players like Emma across all touchpoints.

1. AI-Driven Personalization:
As AI technology advances, casinos will be able to deliver

even more personalized content and offers based on real-time data from multiple platforms. For Emma, this means receiving offers that are perfectly tailored to her preferences, no matter which platform she's using.

2. Augmented Reality and Virtual Reality:
The integration of augmented reality (AR) and virtual reality (VR) into multi-channel strategies will create new opportunities for player engagement. Emma might be able to use AR to explore the casino floor from her mobile app, or participate in virtual slot tournaments from the comfort of her home. These immersive experiences will further blur the lines between physical and digital interactions.

3. More Seamless Cross-Channel Journeys:
The future of multi-channel strategies will see even greater integration between platforms, allowing players like Emma to move fluidly between physical and digital experiences. For instance, Emma might start playing a game on her phone, continue it on her laptop, and finish it in the casino—all without losing progress or rewards.

Key Takeaways

1. **Multi-Channel Strategies Enhance Engagement:**
   By offering seamless engagement across physical, digital, and mobile platforms, casinos can keep players like Emma connected and engaged, regardless of where they are or what device they're using.
2. **Cross-Channel Integration is Essential:**
   A successful multi-channel strategy relies on integrating data, rewards, and experiences across all platforms, ensuring that players enjoy a unified and consistent experience at every touchpoint.
3. **Personalization Increases Engagement and Loyalty:**
   Using data from multiple platforms, casinos can deliver personalized content, offers, and promotions that resonate with players like Emma, making her feel valued and increasing her loyalty.
4. **Real-Time Engagement Keeps Players Connected:**
   Real-time notifications, promotions, and reminders help keep

players engaged across channels, encouraging them to take immediate action and increasing overall player retention.

5. **The Future is AI-Powered and Immersive:**
As AI and emerging technologies like AR and VR evolve, multi-channel strategies will become even more dynamic and personalized, offering players like Emma an immersive, interconnected experience that blends physical and digital gaming.

4.

THE VALUE EQUATION

Maximizing Player Lifetime Value (LTV)

Emma isn't just any player at the casino; she's a regular who visits every few weeks, spends time on both the slot machines and table games, and participates in the loyalty program. While she might not be a high-roller on every visit, over time her contributions to the casino add up to a substantial amount. The casino recognizes Emma's value—not just based on what she spends in a single night, but what she could contribute over months and even years. This is the concept of Player Lifetime Value (LTV), and for the casino, maximizing Emma's LTV is key to building a sustainable, profitable business.

What is LTV

Player Lifetime Value (LTV) is a crucial metric that estimates the total revenue a player like Emma will generate for the casino throughout their entire relationship. Rather than focusing on short-term gains, LTV shifts the focus to the long-term profitability of each player. In Emma's case, her LTV is not determined by her spending during a single visit, but by how often she returns, how much she engages with the casino's offerings, and how loyal she remains over time.

LTV helps casinos make informed decisions about marketing, promotions, and player retention strategies by identifying the players who provide the most value over the long haul. For casinos, maximizing LTV means investing in strategies that keep players like Emma engaged, satisfied, and coming back for more. By increasing Emma's engagement, the casino can not only maximize her immediate spend but also cultivate a long-term relationship that generates continuous revenue.

In this chapter, we will explore strategies for maximizing Player Lifetime Value (LTV), focusing on the factors that drive long-term engagement and loyalty, the role of personalized experiences, and how AI and data analytics are transforming the way casinos build lasting relationships with players like Emma.

The Importance

Maximizing LTV is essential for casinos because it focuses on long-term sustainability rather than short-term wins. While attracting new players is important, retaining existing players and increasing their engagement over time is often more profitable and cost-effective. In fact, research across industries shows that improving customer retention by just 5% can lead to a 25% to 95% increase in profits.

For players like Emma, maximizing LTV means creating an environment where she feels valued and incentivized to keep returning. This involves offering more than just games and promotions—it means crafting a holistic experience that encourages Emma to engage across multiple touchpoints, from the casino floor to the mobile app to online platforms. By focusing on Emma's LTV, the casino can prioritize long-term growth and profitability by deepening its relationship with her.

Key Drivers

Several factors contribute to maximizing player LTV, each of which plays a critical role in building a lasting relationship with players like Emma. These drivers include frequency of visits, player engagement, loyalty program participation, and overall player satisfaction. Let's explore how each of these elements can be leveraged to increase player LTV.

1. Frequency of Visits

The more often a player like Emma visits the casino, the higher her LTV will be. Encouraging frequent visits is essential to maximizing her value, as repeat visits provide more opportunities for engagement and spending. Casinos can use a variety of strategies to increase the frequency of visits, from personalized promotions to exclusive events and ongoing rewards that incentivize return visits.

For Emma, receiving a personalized promotion that aligns with her preferences, such as a weekend offer for extra loyalty points on her favorite slot machine, would encourage her to visit more regularly. By offering frequent rewards for her continued engagement, the casino keeps her motivated to return, ultimately increasing her LTV.

2. Player Engagement Across Channels

Player engagement across channels—whether in the physical casino, on the mobile app, or through online gaming—directly impacts LTV. The more engaged a player is, the more likely they are to spend time and money at the casino. Casinos should aim to create a seamless, multi-channel experience that encourages players like Emma to interact with the brand across all touchpoints.

For instance, Emma could start her gaming experience at the physical casino, earn loyalty points, and then continue playing on the casino's mobile app when she's at home. By providing cross-channel rewards and ensuring that her progress is synchronized across platforms, the casino deepens Emma's engagement and increases her likelihood of spending more over time.

3. Loyalty Program Participation

One of the most powerful tools for maximizing player LTV is the casino's loyalty program. Players who actively participate in loyalty programs tend to have higher LTVs because they are incentivized to return and engage with the casino regularly. Loyalty programs offer tangible rewards, such as free play, dining credits, or exclusive access to events, that encourage continued engagement.

For Emma, the casino's loyalty program allows her to accumulate points based on her activity, which she can redeem for rewards. By offering personalized perks, such as bonuses on her favorite games or exclusive invitations to special events, the casino keeps Emma engaged, increasing her LTV.

Casinos can further optimize their loyalty programs by offering tiered rewards, where players earn progressively better rewards as they reach higher levels of engagement. For Emma, this could mean moving from a basic rewards tier to a premium level where she gains access to more exclusive benefits, further motivating her to stay loyal and engaged with the casino.

4. Personalized Player Experiences

Personalization is a key factor in maximizing LTV because it makes players feel valued and understood. By tailoring offers, promotions, and experiences to match each player's unique preferences and behaviors, casinos can create more meaningful and relevant interactions that increase engagement.

For example, if Emma consistently plays a certain type of slot machine, the casino could send her promotions specifically for that game, or offer her free spins to try out a new, similar machine. This level of personalization makes Emma feel like the casino is catering to her interests, increasing her likelihood of continuing to engage with the brand.

AI and data analytics play a crucial role in delivering these personalized experiences. By analyzing Emma's gameplay data, spending habits, and preferences, the casino can craft promotions and offers that resonate with her on a personal level, enhancing her overall experience and increasing her LTV.

AI and Predictive Analytics to Maximize LTV

Artificial intelligence (AI) and predictive analytics have become essential tools for maximizing player LTV, allowing casinos to analyze vast amounts of player data and predict future behavior. By leveraging AI, casinos can create highly personalized player journeys that increase engagement, retention, and long-term profitability.

1. Predicting Player Behavior: AI-powered predictive analytics can forecast Emma's future behavior based on her past actions. For instance, the casino can predict when Emma is most likely to visit, what games she will play, and how much she is likely to spend. By anticipating her needs, the casino can deliver timely, relevant promotions that encourage her to return more frequently.

2. Real-Time Personalization: AI allows casinos to personalize Emma's experience in real time, adjusting offers and promotions based on her current behavior. If Emma has been playing for a while without winning, the AI system might offer her a bonus round or free spins to keep her engaged. Conversely, if Emma is on a winning streak, the AI could suggest higher-stakes games to encourage further play.

3. Optimizing Marketing Efforts: Predictive analytics also help casinos optimize their marketing efforts by identifying which players are most likely to respond to specific promotions. For Emma, this means that she receives offers that are tailored to her preferences, increasing the likelihood that she will engage with the promotion. AI-driven marketing campaigns are more effective because they target the right players with the right message at the right time, increasing engagement and driving long-term value.

Strategies to Maximize LTV

To maximize player LTV, casinos must implement strategies that focus on long-term engagement and relationship building. Some of the most effective strategies include:

1. Personalize Every Interaction: Every interaction with a player like Emma should feel personalized and tailored to her preferences. This includes sending targeted promotions, recommending games based on her interests, and offering rewards that align with her gaming habits.

2. Encourage Multi-Channel Engagement: Encourage players to engage with the casino across multiple channels by offering cross-channel rewards and promotions. For Emma, this means that her progress, rewards, and promotions are accessible no matter where she plays, whether it's on the casino floor, through the mobile app, or online.

3. Implement Tiered Loyalty Programs: Offer tiered loyalty programs that provide progressively better rewards as players reach higher levels of engagement. By incentivizing Emma to reach higher tiers, the casino encourages her to return more often and engage with the brand in deeper ways.

4. Use AI to Predict and Personalize: Leverage AI and predictive analytics to anticipate player behavior and deliver real-time personalized experiences. AI can help casinos deliver the right offers to Emma at the right time, increasing her engagement and maximizing her LTV.

Sample LTV Calculation Formulas

Casinos use various formulas to calculate **Lifetime Value (LTV)**, tailored to their unique business models and types of player interactions. These formulas consider factors such as player spending, retention rates, and frequency of visits. Here are some of the most common LTV calculation methods used in the casino industry:

1. Basic LTV Formula

A simplified approach that estimates the average revenue generated by a player during their engagement period.

$$LTV = ARPU \times AL$$

Where:

- ARPU = Average Revenue Per User (total revenue divided by the number of active users)
- AL = Average Lifespan (the average length of time a player remains active)

Use Case: Suitable for basic estimations when detailed data is limited.

2. Detailed LTV Formula with Gross Margin

An advanced formula that accounts for the gross margin to reflect profitability rather than just revenue.

$$LTV = (ARPU \times AL \times GM) - CAC$$

Where:

- GM = Gross Margin (percentage of revenue remaining after deducting direct costs)
- CAC = Customer Acquisition Cost (the cost of acquiring each new player)

Use Case: Useful for more precise financial analysis, especially when evaluating the profitability of player acquisition campaigns.

3. Retention-Based LTV Formula

Focuses on retention rates to model how long players are expected to remain active and contribute revenue.

$$LTV = (ARPU \times r) / (1 + d - r)$$

Where:

- r = Retention Rate (the percentage of players retained over a specific period)
- d = Discount Rate (accounts for the time value of money)

Use Case: This formula is ideal for casinos focusing on long-term player retention strategies and accounts for customer churn.

4. Cohort Analysis LTV

Calculates LTV based on the behavior of specific player cohorts to provide a more segmented view.

$$LTV_{cohort} = \sum_{t=1}^{n} \frac{R_t}{(1+d)^t}$$

Where:

- R_t = Revenue generated by the cohort at time t
- d = Discount Rate
- n = Number of periods observed

Use Case: This method is useful for understanding the value of different types of players (e.g., high rollers vs. casual players) and tailoring marketing efforts accordingly.

5. Customer Engagement Score LTV

Incorporates a weighted engagement score to predict LTV based on player interaction levels.

$$LTV = \left(\frac{\sum_{i=1}^{n} E_i \times S_i}{n}\right) \times AL$$

Where:

- E_i = Engagement factor for each interaction type (e.g., play frequency, session length)
- S_i = Score assigned to each type of interaction based on its impact on revenue
- n = Total number of interactions measured

Use Case: This approach is valuable for casinos that leverage comprehensive player data to assess LTV based on engagement patterns.

6. Predictive LTV Using Machine Learning

Machine learning models, such as **regression analysis** or **neural networks**, can predict LTV by analyzing player behavior and historical data.

Model Example:

$$LTV_{predicted} = f(X)$$

Where:

- $f(X)$ = Machine learning model trained on variables X (e.g., frequency of visits, average spend per session, player demographic data)

Use Case: Highly effective for casinos with large data sets and the capability to build predictive models that provide dynamic, real-time LTV calculations.

7. Churn-Adjusted LTV

Accounts for the rate at which players stop engaging with the casino, adjusting LTV calculations to reflect churn more accurately.

$$LTV = \frac{ARPU}{ChurnRate}$$

Where:

Churn Rate = The percentage of players lost over a specific period.

Use Case: Ideal for casinos with high player turnover, as it highlights the importance of player retention in long-term revenue.

8. Net Promoter Score (NPS)-Enhanced LTV

Adjusts LTV based on the Net Promoter Score (NPS), which measures customer loyalty and likelihood of recommending the casino.

$$LTV = ARPU \times AL \times \left(1 + \frac{NPS}{100}\right)$$

Where:

NPS = Net Promoter Score (ranges from -100 to +100)

Use Case: This method benefits casinos by focusing on player advocacy and word-of-mouth marketing impact.

Pros and Cons of Different LTV Models

Model	Pros	Cons
Basic LTV	Easy to calculate and understand	Limited insights; does not account for acquisition costs or retention
Detailed LTV with Gross Margin	Reflects profitability, not just revenue	Requires accurate cost and margin data
Retention-Based LTV	Accounts for churn and retention	More complex; sensitive to retention and discount rate assumptions
Cohort Analysis LTV	Provides segmented insights	Requires more data and cohort tracking
Customer Engagement Score LTV	Integrates engagement metrics for deeper analysis	Time-consuming to implement and measure
Predictive LTV (ML-based)	Highly accurate and dynamic	Requires large data sets and advanced analytics tools
Churn-Adjusted LTV	Highlights the impact of churn	Simplistic; may overlook other key variables
NPS-Enhanced LTV	Incorporates loyalty and advocacy	Relies on subjective survey data

Choosing the right LTV model may depend on the data available, the complexity your team can handle, and your casino's specific business goals.

Key Takeaways

1. **Player Lifetime Value (LTV) Focuses on Long-Term Profitability:**
   LTV measures the total revenue a player like Emma generates

over time, making it essential for casinos to prioritize long-term engagement and retention strategies rather than focusing solely on short-term gains.

2. **Engagement, Frequency of Visits, and Loyalty Drive LTV:**
Increasing the frequency of visits, cross-channel engagement, and loyalty program participation are key factors that contribute to maximizing LTV. Players who engage frequently across multiple platforms provide greater value to the casino over time.

3. **Personalization is Key to Building Lasting Relationships:**
Personalized experiences tailored to individual player preferences increase engagement, satisfaction, and loyalty. Casinos can use data analytics and AI to craft personalized offers that resonate with players like Emma, driving long-term value.

4. **AI and Predictive Analytics Optimize LTV Strategies:**
AI and predictive analytics allow casinos to predict player behavior, deliver real-time personalized experiences, and optimize marketing efforts. By leveraging these tools, casinos can engage players more effectively and increase their overall LTV.

5. **LTV Maximization Requires a Long-Term, Relationship-Focused Approach:**
To maximize LTV, casinos must focus on building long-term relationships with players, encouraging multi-channel engagement, and delivering consistent value through personalized experiences and rewards.

5.

IN THE MOMENT

How Real-Time Marketing Changes Loyalty

As Emma walks into the casino, her phone vibrates with a notification: "Welcome back, Emma! Play any slot machine in the next 30 minutes and earn double loyalty points." She feels excited and appreciated—the casino knows exactly when and how to engage her. A few hours later, after playing her favorite slot machine, another message pops up: "You've unlocked a bonus for free spins on your favorite game! Head over to the promotion kiosk to claim your prize." Emma grins—this is just the immediate gratification that makes her feel connected to the casino. Real-time marketing has transformed her experience, making every moment count.

The Power of Real-Time Marketing

In an era where immediate gratification drives consumer behavior, **real-time marketing** has emerged as a powerful tool for casinos to deepen player engagement and loyalty. Real-time marketing refers to the use of data and technology to deliver personalized offers, rewards, and experiences to players like Emma in real-time, based on their current activity or behavior. Rather than waiting for promotions or rewards to come later, players receive instant incentives that encourage them to take immediate action.

Real-time marketing creates a dynamic experience for players like Emma that feels responsive and personalized. Whether it's a bonus for playing a specific game, an offer based on her current spending habits, or an exclusive promotion triggered by her arrival at the casino, real-time marketing ensures that Emma's engagement with the casino is continuously rewarded. This immediate interaction enhances her sense of loyalty, as she feels that the casino is actively paying attention to her behavior and offering timely rewards that align with her interests.

For casinos, real-time marketing offers a way to influence player behavior as it happens, driving deeper engagement and increasing player lifetime value (LTV). By offering promotions at the right moment, casinos can encourage players to stay longer, spend more, and return more frequently. This chapter explores how real-time marketing is transforming player loyalty, the technologies behind it, and the strategies casinos can use to maximize its impact.

How Real-Time Marketing Changes Loyalty

Real-time marketing is changing the way casinos build loyalty by shifting the focus from delayed gratification to immediate rewards. Traditional loyalty programs typically rely on players accumulating points or rewards over time, which can create a sense of detachment between a player's current actions and future benefits. Real-time marketing, on the other hand, closes this gap by offering instant rewards that are directly tied to a player's in-the-moment behavior.

For Emma, this means that her actions at the casino—whether it's choosing a specific game, playing for a set amount of time, or reaching a certain spending threshold—are immediately recognized and rewarded. This creates a sense of excitement and urgency, as Emma knows that her efforts are constantly being monitored and rewarded in real time. The result is a stronger emotional connection to the casino, as Emma feels that her loyalty is being acknowledged and appreciated in the moment, rather than at a distant future date.

This immediate feedback loop strengthens loyalty by making players feel valued and engaged throughout their visit. Rather than waiting to see the benefits of their loyalty at a later date, players like Emma experience the rewards of their loyalty in real time, which encourages them to stay longer and return more frequently.

Key Benefits

Real-time marketing offers several key benefits for casinos, each of which plays an important role in building and maintaining player loyalty. Casinos can deepen their connection with players like Emma by offering personalized, immediate incentives and encouraging long-term engagement.

1. Immediate Rewards for Immediate Actions: One of the most significant benefits of real-time marketing is the ability to offer instant rewards based on a player's current actions. For Emma, this might mean receiving free spins after reaching a certain number of spins on a slot machine or earning a bonus for switching from a table game to a slot game within a set period. These immediate rewards make Emma feel that her actions are being recognized and valued, encouraging her to keep playing and engage more deeply with the casino.

Immediate rewards also create a sense of excitement and urgency, as Emma knows that she could unlock special bonuses or promotions at any moment based on her current behavior. This keeps her engaged and motivated to continue playing, as the next reward could be just a few actions away.

2. Personalized Offers Based on Real-Time Data: Real-time marketing allows casinos to deliver personalized offers that are tailored to each player's current behavior. By analyzing Emma's activity in real time—whether it's the games she's playing, how much she's spending, or how long she's been at the casino—the casino can deliver targeted promotions that resonate with her preferences.

For example, if Emma tends to play a specific type of slot machine, the casino could offer her a bonus for trying a new game that has similar features. If she's been playing for a while without winning, the casino might offer her free spins to keep her engaged. This level of personalization makes Emma feel that the casino is actively paying attention to her preferences, which strengthens her sense of loyalty.

3. Encouraging In-the-Moment Behavior Changes: Real-time marketing can also be used to influence player behavior in the moment, encouraging players to try new games, spend more time at the casino, or explore different areas of the casino floor. For example, if Emma has been playing at a slot machine for an extended period, the casino might offer her a promotion to try a table game or participate in a tournament. By incentivizing her to switch games or explore new experiences, the casino keeps her engaged and prevents boredom.

Additionally, real-time marketing can be used to drive spending. If Emma is close to reaching a spending threshold that would unlock a special reward, the casino could send her a notification encouraging her to make one more bet to claim the prize. This strategy not only increases immediate spending but also enhances Emma's experience by giving her a sense of accomplishment.

4. Driving Longer Play Sessions and Increased Spending: Real-time marketing encourages players to stay longer and spend more by offering continuous rewards and incentives throughout their visit. For Emma, this might mean receiving a promotion for free drinks after playing for a set amount of time, or earning bonus loyalty points for participating in multiple games during her visit.

These ongoing rewards create a sense of progression, as Emma feels that every action she takes is moving her closer to unlocking more bonuses or promotions. This keeps her motivated to continue playing, increasing both her playtime and overall spending. For the casino, this translates into higher revenue and greater player retention, as Emma is more likely to return for future visits to unlock more rewards.

5. Enhancing the Overall Player Experience: Perhaps the most important benefit of real-time marketing is its ability to enhance the overall player experience. By delivering personalized, timely rewards and promotions, casinos can create an engaging, dynamic environment that keeps players like Emma excited and motivated throughout their visit.

For Emma, real-time marketing makes every moment at the casino feel rewarding, as she knows that her actions are being recognized and appreciated in real time. This creates a positive emotional connection to the casino, which strengthens her loyalty and makes her more likely to return in the future.

Technologies

The success of real-time marketing relies on advanced technologies that allow casinos to analyze player behavior in real time, deliver personalized promotions, and track player engagement across multiple channels. Some of the key technologies behind real-time marketing include:

1. Artificial Intelligence (AI) and Machine Learning: AI and machine learning play a critical role in real-time marketing by analyzing vast amounts of player data and predicting which promotions or rewards are most likely to resonate with each player. For Emma, this means that the casino's AI system is constantly monitoring her behavior—whether she's playing slot machines, participating in table games, or interacting with the mobile app—and using this data to deliver personalized, real-time offers that align with her preferences.

AI also enables casinos to predict player behavior and deliver timely promotions that encourage immediate action. For example, if Emma is close to reaching a spending threshold, the AI system might offer her a promotion that incentivizes her to keep playing and unlock a reward.

2. Real-Time Data Analytics: Real-time data analytics allows casinos to track player activity in real time and use this data to deliver immediate rewards and promotions. For Emma, this means that every spin, bet, or game she plays is being tracked, and the casino can instantly recognize when she reaches certain milestones or qualifies for a promotion.

By analyzing this data in real time, casinos can deliver rewards that are perfectly timed to match Emma's current behavior. For instance, if she's been playing for a while without winning, the casino could offer her a bonus or free spins to keep her engaged. This real-time analysis ensures that promotions are relevant and timely, increasing their effectiveness.

3. Mobile and Push Notifications: Mobile technology plays a key role in real-time marketing by allowing casinos to engage players like Emma through push notifications,

SMS, or in-app messages. As soon as Emma enters the casino or engages with the mobile app, the casino can send her personalized offers and promotions that encourage her to take immediate action.

For example, if Emma enters the casino and hasn't played in a while, she might receive a push notification offering her a promotion for free spins if she plays within the next 30 minutes. This immediate engagement encourages Emma to start playing right away, increasing her overall engagement and loyalty.

4. Cross-Channel Integration: To deliver a seamless real-time marketing experience, casinos must integrate their systems across multiple channels, including the physical casino, mobile app, and online platforms. For Emma, this means that her activity is tracked and rewarded across all channels, whether she's playing in the casino or using the app at home.

Cross-channel integration ensures that Emma's real-time promotions and rewards are synchronized across platforms, creating a cohesive experience that encourages her to engage with the casino no matter where she is.

Strategies

To fully leverage the power of real-time marketing, casinos must implement strategies that focus on delivering personalized, timely offers that resonate with players like Emma. Here are some key strategies for maximizing the impact of real-time marketing:

1. Deliver Immediate Rewards for Specific Actions: Offer immediate rewards based on specific actions that players take in the moment. For example, if Emma reaches a certain number of spins on a slot machine, she could receive a bonus for free spins or a loyalty point multiplier. These immediate rewards create a sense of progression and encourage Emma to keep playing.

2. Use Real-Time Promotions to Drive Behavior Changes: Use real-time promotions to influence player behavior in the moment. For example, if Emma has been playing slot machines for an extended period, the casino could offer her a promotion to try a new table game or participate in a tournament. By incentivizing Emma to switch games or explore different areas of the casino, the casino keeps her engaged and interested.

3. Send Personalized Notifications Based on Player Activity: Use mobile push notifications or in-app messages to send personalized offers based on Emma's current activity. For example, if Emma hasn't played in a while, the casino could send her a promotion offering double loyalty points for her next visit. These timely notifications encourage immediate action and keep Emma engaged with the casino.

4. Reward Cross-Channel Engagement: Encourage players to engage across multiple channels by offering rewards for cross-channel activity. For example, if Emma plays a game on the mobile app, she could receive a promotion to continue playing in the physical casino. This cross-channel engagement keeps Emma connected to the casino across all touchpoints, increasing her overall loyalty and LTV.

The Future

As technology continues to advance, the future of real-time marketing will become even more personalized, dynamic, and immersive. Emerging technologies like artificial intelligence, machine learning, and augmented reality will further enhance the ability of casinos to deliver real-time rewards and promotions that resonate with players like Emma.

1. Hyper-Personalization: In the future, real-time marketing will become even more personalized, with AI systems delivering offers that are perfectly tailored to each player's unique preferences and behavior. For Emma, this means that every promotion she receives will feel custom-made for her, increasing her engagement and loyalty.

2. Augmented Reality and Gamification: Augmented reality (AR) and gamification will play a larger role in real-time marketing, allowing casinos to create interactive, gamified experiences that engage players in the moment. For example, Emma could participate in an AR treasure hunt within the casino, where she earns rewards for completing certain challenges or reaching specific milestones.

3. Real-Time Feedback Loops: As real-time marketing continues to evolve, casinos will use feedback loops to create continuous engagement throughout the player journey. For Emma, this means that every action she takes—whether it's playing a game, interacting with the app, or making a purchase—will trigger immediate feedback in the form of rewards or promotions, keeping her engaged and motivated at all times.

Sample Marketing Formulas

Real-time marketing in a casino environment is fairly complex and leverages various data sets and player interactions to deliver timely, personalized campaigns and offers. Below are some formulas for real-time marketing strategies in casinos designed to optimize player engagement and maximize ROI:

1. Real-Time Engagement Score (RTES)

A formula to assess the likelihood of a player engaging with a real-time marketing offer.

RTES = (F × W1) + (S × W2) + (R × W3)

Where:

- F = Frequency of visits (number of times a player visits the casino in a specific period)
- S = Session length (average time spent per visit)
- R = Real-time interactions (e.g., clicks on promotions or app usage during the visit)
- W1, W2, W3 = Weights assigned to each factor based on their importance

Application: This score helps casinos identify which players are most likely to engage with real-time offers, allowing them to tailor campaigns accordingly.

2. Real-Time Offer Optimization (RTO)

Calculates the best offer to present to a player based on their recent behavior and profile.

$$RTO = \frac{E_p \times C_v}{T_r}$$

Where:

- E_p = Estimated profitability of the offer (e.g., projected spend from the player)
- C_v = Current value of the player (average revenue generated by the player per visit)
- T_r = Time since the last relevant interaction (measured in minutes or hours)

Application: Casinos can use this formula to prioritize which offers are sent to which players, ensuring maximum ROI and real-time engagement.

3. Predictive Engagement Probability (PEP)

Estimates the probability that a player will engage with a marketing message based on historical and real-time data.

$$PEP = \sigma(B + \beta_1 X_1 + \beta_2 X_2 + \beta_3 X_3)$$

Where:

- σ = Sigmoid function for probability output (bounded between 0 and 1)
- B = Bias term
- X_1 = Real-time player interaction data (e.g., time spent on a specific game)
- X_2 = Player preferences (e.g., game types or dining options)
- X_3 = Current promotions or offers viewed by the player
- $\beta_1, \beta_2, \beta_3$ = Coefficients indicating the weight of each factor

Application: This predictive model helps determine how likely a player is to engage with an offer, informing when and what type of real-time message to deliver.

4. Dynamic Offer Value (DOV)

Calculates the value of an offer based on the player's potential response and the casino's current capacity or situation.

$$DOV = \frac{P_r \times L_p}{A_c + E}$$

Where:

- P_r = Probability of player redemption (based on past behavior)
- L_p = Potential lifetime profit from the player
- A_c = Current capacity of the casino (e.g., available spots in a promotion or event)
- E = External factors (e.g., time of day, day of the week, special events)

Application: Useful for determining the optimal value of offers during peak and off-peak times, allowing casinos to adjust promotions in real-time.

5. Real-Time Conversion Rate (RTCR)

Measures the effectiveness of real-time marketing campaigns in converting player interactions into engagement.

$$RTCR = \frac{N_e}{N_t}$$

Where:

- N_e = Number of players who engaged with the real-time offer (clicked, redeemed, etc.)
- N_t = Total number of players who received the offer

Application: This metric provides immediate feedback on the success of real-time campaigns, helping marketers optimize strategies on the fly.

6. Immediate Return on Marketing Spend (iROMS)

Calculates the return on investment for real-time marketing efforts over a short period.

$$iROMS = \frac{R_m - C_m}{C_m} \times 100$$

Where:

- R_m = Revenue generated from the real-time marketing campaign
- C_m = Cost of executing the marketing campaign

Application: Provides casinos with a quick snapshot of the financial success of their real-time marketing initiatives.

7. Player Excitement Index (PEI)

A measure to evaluate how engaged players are with real-time events, offers, or interactions.

$$PEI = \frac{(I_c \times P_f) + A_s}{2}$$

Where:

- I_c = Interaction count (e.g., number of interactions with the real-time campaign)
- P_f = Player feedback score (obtained through quick surveys or interactions)
- A_s = Average session duration after interaction

Application: Helps gauge the excitement levels of players to refine future campaigns and real-time offers.

8. Real-Time Engagement ROI (RT-ROI)

Evaluates the total return on real-time marketing investments over an extended period (daily/weekly/monthly).

$$RT - ROI = \frac{(T_r - T_c) \times L}{T_c}$$

Where:

- T_r = Total revenue generated by real-time campaigns
- T_c = Total cost of the campaigns
- L = Average lifetime of engaged players from the campaign

Application: Provides a comprehensive look at how real-time strategies impact the overall casino profitability.

These formulas may allow casinos to use data with AI to optimize real-time marketing efforts. Casinos can better manage marketing dollars by calculating potential responses, campaign effectiveness, and ROI and achieve higher player engagement and retention.

Key Takeaways

1. **Real-Time Marketing Delivers Immediate Rewards:** Real-time marketing allows casinos to offer instant rewards based on player actions, creating a sense of urgency and excitement that keeps players like Emma engaged.
2. **Personalized Offers Deepen Engagement:** By using real-time data analytics and AI, casinos can deliver personalized promotions that resonate with each player's unique preferences, increasing their loyalty and satisfaction.
3. **Real-Time Promotions Influence Behavior:** Real-time marketing can be used to influence player behavior in the moment, encouraging players to try new games, spend more time at the casino, or increase their spending.
4. **Mobile and Push Notifications Drive Immediate Action:** Mobile technology allows casinos to engage players through real-time push notifications and in-app messages, encouraging immediate action and increasing player engagement.

5. **The Future of Real-Time Marketing is Hyper-Personalized and Immersive:**
   As technology continues to evolve, real-time marketing will become even more personalized and interactive, with AI.

6.

LOYALTY UNLOCKED

Revealing the True Potential of Loyalty Programs

As Emma sits at her favorite slot machine, she glances at her phone to check her loyalty points. The app shows that she's close to unlocking a special bonus—a free dinner at the casino's five-star restaurant. Emma smiles, knowing that every game she plays, every dollar she spends, is getting her closer to another reward. For her, the loyalty program isn't just an afterthought—it's an integral part of her experience at the casino. Each reward feels like a tangible acknowledgment of her commitment, and it keeps her coming back, eager to see what new benefits she can unlock. Emma's loyalty isn't just a result of good games or fancy promotions; it's the casino's well-crafted loyalty program that consistently rewards her engagement, ensuring she feels valued at every step.

The Power

Loyalty programs are one of the most powerful tools casinos have to cultivate long-term relationships with players like Emma. A well-designed loyalty program goes beyond simple rewards—it's a strategic approach to building player loyalty, increasing engagement, and maximizing lifetime value (LTV). For players, loyalty programs provide an incentive to return, offering tangible rewards and recognition for their ongoing participation. For casinos, loyalty programs drive repeat visits, boost player retention, and create a more personalized gaming experience that keeps players coming back for more.

In this chapter, we'll explore the true potential of loyalty programs, focusing on how they enhance player engagement, increase retention, and ultimately transform casual players into loyal patrons. From tiered rewards systems to personalized experiences, loyalty programs are a key driver of long-term success in the casino industry. For players like Emma, a well-structured loyalty program provides a sense of progression, rewards for consistent play, and personalized offers that make every visit more enjoyable.

The Role

At the heart of every loyalty program is the idea of **rewarding engagement**. By offering points, bonuses, and exclusive perks for every action a player takes—whether it's playing a game, spending money, or visiting the casino—loyalty programs encourage players like Emma to stay engaged and return for more.

For Emma, the loyalty program creates a clear connection between her actions and rewards. Every time she plays her favorite slot machine, she earns points that can be redeemed for bonuses, free play, or other rewards. This immediate gratification keeps her motivated to continue playing, knowing that every dollar she spends brings her closer to unlocking something valuable. As Emma progresses through the loyalty program, the rewards become more enticing, encouraging her to visit more frequently and spend more time at the casino.

Loyalty programs don't just reward individual actions—they create a sense of belonging and recognition. For Emma, the loyalty program makes her feel like a valued member of the community rather than just another player. The more she engages with the program, the more she feels that the casino recognizes her loyalty and is actively rewarding her for it.

The True Potential

Loyalty programs have the potential to do more than just reward players—they can drive deeper engagement, increase player retention, and create long-term value for both the player and the casino. A well-designed loyalty program taps into key psychological drivers like the need for recognition, achievement, and progression, turning casual players like Emma into loyal patrons who are invested in the brand.

1. Fostering a Sense of Progression and Achievement

One of the most powerful aspects of a loyalty program is its ability to create a sense of **progression**. For Emma, every visit to the casino is another step toward unlocking a new reward or reaching a higher tier in the program. The loyalty program gives her a clear path to follow, with each action—whether it's playing a game, earning points, or spending money—moving her closer to the next milestone.

This sense of progression keeps Emma motivated and engaged. She isn't just playing for the sake of playing; she's working toward tangible goals, such as unlocking exclusive rewards or moving into a higher loyalty tier where she can access even better benefits. The loyalty program turns each visit into a journey, with Emma constantly striving to reach the next level.

Tiered loyalty programs, where players move through different levels based on their engagement, are particularly effective at fostering this sense of achievement. As Emma progresses through the program, she unlocks better rewards, such as higher point multipliers, access to VIP events, or personalized promotions. This progression makes Emma feel like she's being recognized and rewarded for her loyalty, encouraging her to stay engaged and continue playing.

2. Offering Personalized Rewards and Experiences

Loyalty programs unlock their full potential when they deliver **personalized rewards and experiences** tailored to each player's preferences. For Emma, this means receiving rewards that align with her favorite games, spending habits, and interests, rather than generic promotions that don't resonate with her.

Personalization is key to creating a loyalty program that feels relevant and engaging. By analyzing Emma's behavior, the casino can tailor rewards and promotions that speak directly to her interests. For example, if Emma frequently plays high-volatility slot machines, the casino might offer her free spins on a new, high-volatility game. If she tends to visit on Friday nights, the casino could send her a personalized promotion offering double loyalty points for playing on a Friday evening.

These personalized rewards make Emma feel like the casino understands her preferences and is actively working to enhance her experience. By tailoring the loyalty program to her unique habits, the casino strengthens her emotional connection to the brand, increasing her likelihood of returning and staying loyal.

3. Encouraging Repeat Visits

One of the main goals of any loyalty program is to encourage **repeat visits**. By offering ongoing rewards, promotions, and bonuses, loyalty programs give players like Emma a reason to return to the casino regularly. The more Emma engages with the program, the more rewards she unlocks, creating a cycle of engagement that keeps her coming back for more.

For example, Emma might receive a promotion offering her extra loyalty points if she visits twice in the same week. Or she could be offered a special bonus for playing on consecutive days. These promotions incentivize Emma to visit the casino more frequently, knowing that each visit brings her closer to unlocking additional rewards.

Loyalty programs can also use time-limited offers to create a sense of urgency, encouraging Emma to visit sooner rather than later. For instance, the casino could send her a personalized offer for free play that's only available if she visits within the next 48 hours. These time-sensitive promotions tap into Emma's fear of missing out, increasing the likelihood that she'll return to the casino sooner.

4. Strengthening Emotional Loyalty

While loyalty programs often focus on **transactional loyalty** (rewarding players for their spending or participation), they can also play a key role in building **emotional loyalty**. Emotional loyalty goes beyond points and rewards—it's about creating a deep, emotional connection between the player and the brand, where the player feels valued, appreciated, and emotionally invested in the casino.

For Emma, the loyalty program does more than just offer rewards—it makes her feel like a valued part of the casino's community. She knows that the more she engages with the program, the more the casino recognizes her loyalty. By offering personalized rewards, VIP experiences, and exclusive perks, the casino creates an environment where Emma feels appreciated and valued, strengthening her emotional attachment to the brand.

Emotional loyalty is particularly important for long-term player retention. When players like Emma feel emotionally connected to a casino, they are more likely to stay loyal even when competing offers or promotions are available elsewhere. By fostering emotional loyalty, the casino can build long-lasting relationships with players, driving higher lifetime value (LTV) and long-term engagement.

5. Driving Long-Term Engagement

Loyalty programs are a key driver of **long-term engagement**, turning one-time players into repeat visitors who are invested in the casino's brand. For Emma, the loyalty program keeps her engaged over time by offering new rewards, exclusive promotions, and ongoing incentives that encourage her to return regularly.

As Emma progresses through the loyalty program, the casino can continue to introduce new elements that keep her engaged, such as special events, personalized bonuses, or access to VIP experiences. By constantly evolving the program and offering fresh incentives, the casino ensures that Emma's interest doesn't wane, keeping her connected to the brand for the long term.

Long-term engagement is critical for maximizing player lifetime value (LTV). By keeping players like Emma engaged over an extended period, the casino can drive higher revenue, increase retention, and create a more loyal player base.

Strategies for Unlocking

To fully unlock the potential of loyalty programs, casinos must focus on delivering personalized, meaningful experiences that keep players like Emma engaged and motivated. Here are some key strategies for creating a loyalty program that drives long-term player engagement:

1. Implement Tiered Loyalty Programs

Tiered loyalty programs are highly effective at fostering a sense of progression and achievement. By offering different levels of rewards based on a player's engagement, casinos can incentivize players like Emma to strive for higher tiers, where the benefits become increasingly valuable.

For example, Emma might start in a basic tier where she earns points for every dollar spent, but as she engages more with the casino, she could progress to higher tiers that offer enhanced rewards, such as point multipliers, free play, or access to exclusive events. This progression creates a sense of achievement and encourages Emma to stay loyal to the casino.

2. Offer Personalized Rewards and Promotions

Personalization is key to creating a loyalty program that resonates with players. Casinos can tailor rewards and promotions to each player's preferences and behavior by analyzing player data. For Emma, this means receiving rewards that align with her favorite games, visit patterns, and spending habits rather than generic offers.

For example, if Emma consistently plays a certain type of slot machine, the casino could offer her free spins on similar games or bonus points for reaching certain milestones. This level of personalization makes Emma feel valued and increases her engagement with the loyalty program.

3. Create Time-Limited Offers to Encourage Immediate Action

Time-sensitive promotions can create a sense of urgency and encourage players to take immediate action. By offering limited-time rewards that expire within a certain period, casinos can drive more frequent visits and increase player engagement.

For instance, Emma might receive a promotion offering her double loyalty points if she visits the casino within the next 48 hours. These time-sensitive offers tap into her desire for immediate rewards, motivating her to return to the casino sooner rather than later.

4. Reward Cross-Channel Engagement

To maximize engagement, loyalty programs should reward players for engaging with the casino across multiple channels, whether it's the physical casino, mobile app, or online platform. For Emma, this means that her loyalty points and rewards are synchronized across all channels, allowing her to accumulate points no matter where she plays.

By offering cross-channel rewards, the casino encourages Emma to engage with the brand in multiple ways, deepening her connection to the loyalty program and increasing her overall engagement.

5. Offer Exclusive Experiences for High-Tier Players

To strengthen emotional loyalty, casinos should offer exclusive experiences for high-tier players, such as access to VIP events, personalized services, or behind-the-scenes tours. These experiences create a sense of exclusivity and make players like Emma feel valued and appreciated.

For Emma, being invited to a VIP event or receiving a personalized offer from the casino makes her feel like a special, valued guest. These exclusive experiences enhance her emotional connection to the brand, increasing her loyalty and long-term engagement.

The Future of Loyalty Programs

As technology continues to evolve, the future of loyalty programs will become even more personalized, dynamic, and immersive. Emerging technologies like artificial intelligence (AI), augmented reality (AR), and predictive analytics will play a key role in shaping the next generation of loyalty programs.

1. AI-Driven Personalization: AI will enable casinos to deliver even more personalized rewards and experiences based on real-time player data. For Emma, this means receiving offers that are perfectly tailored to her preferences, behaviors, and current activity, increasing her engagement with the program.

2. Augmented Reality (AR) Experiences: AR will play a role in creating immersive loyalty experiences, where players like Emma can unlock rewards or participate in interactive promotions through augmented reality. For example, Emma could use her phone to scan a QR code in the casino and participate in an AR-based treasure hunt that offers exclusive rewards.

3. Predictive Analytics for Tailored Promotions:
Predictive analytics will allow casinos to anticipate player behavior and deliver timely promotions that align with a player's preferences and habits. For Emma, this means receiving promotions that are perfectly timed to encourage her to visit the casino at the right moment, increasing her likelihood of returning.

Sample Predictive Loyalty Formulas

Predictive loyalty calculation involves analyzing multiple factors to estimate how likely a player is to remain loyal and engaged across different casino channels. Below are proposed formulas tailored to various aspects of player activity and behavior in a casino setting:

1. Overall Loyalty Score (OLS)

Calculates the overall loyalty of a player by aggregating their activity across various channels.

$$OLS = \frac{\sum_{i=1}^{n}(A_i \times W_i)}{n}$$

Where:
- A_i = Activity score for each channel (slot games, table games, restaurants, etc.)
- W_i = Weight assigned to each channel based on its importance in driving loyalty
- n = Number of channels considered

Application: This formula provides a weighted loyalty score to identify which channels contribute most to player loyalty.

2. Predictive Slot Game Loyalty (PSGL)

Estimates loyalty specifically for slot game players based on their playing patterns and preferences.

$$PSGL = \left(\frac{F_s \times S_d \times R_s}{T_p}\right) \times C_g$$

Where:

- F_s = Frequency of slot game sessions
- S_d = Average session duration for slot games
- R_s = Revenue generated per session
- T_p = Total playing time across all slot games
- C_g = Custom coefficient for player demographics and preferences

Application: Helps identify high-value slot players and potential loyalty boosters for this specific channel.

3. Table Game Engagement Index (TGEI)

Calculates loyalty based on a player's interaction with table games

$$TGEI = \frac{N_t \times B_a}{A_t + D_c}$$

Where:

- N_t = Number of table game sessions
- B_a = Average bet amount
- A_t = Total time spent on table games
- D_c = Discount factor for low-frequency players

Application: Useful for assessing loyalty among table game enthusiasts and informing targeted campaigns for this group.

4. Interest-Based Loyalty (IBL)

Estimates loyalty based on a player's participation in non-gambling activities such as golf, events, and travel etc.

$$IBL = \sum_{j=1}^{m} \frac{(P_j \times E_j \times T_j)}{D}$$

Where:

- P_j = Participation frequency in non-gambling activities
- E_j = Engagement level (measured by spending or time)
- T_j = Time spent on each activity
- D = Demographic factor that influences activity preference
- m = Number of non-gambling activities considered

Application: Helps in understanding how auxiliary offerings contribute to overall player loyalty.

5. Restaurant and Shopping Loyalty (RSL)

Calculates loyalty based on dining and retail spending.

$$RSL = \left(\frac{S_r \times F_r}{T_r}\right) \times L_f$$

Where:

- S_r = Total spending in restaurants and shops
- F_r = Frequency of visits to these venues
- T_r = Total time spent in these activities
- L_f = Loyalty factor adjusted for repeat visits and special events

Application: This score helps identify high-value customers who contribute through dining and retail purchases.

6. Event Participation Loyalty (EPL)

Measures loyalty based on participation in events hosted by the casino.

$$EPL = \frac{(E_p \times A_e \times R_e)}{T_e}$$

Where:

- E_p = Number of events participated in
- A_e = Average attendance duration
- R_e = Revenue generated from event participation
- T_e = Total time available for event participation

Application: Allows casinos to understand the impact of events on player loyalty and tailor event marketing strategies.

7. Demographic-Adjusted Loyalty Score (DALS)

Accounts for demographic factors such as age, gender, and income to predict loyalty.

$$DALS = \left(\frac{OLS \times D_w}{P_a}\right)$$

Where:

- OLS = Overall Loyalty Score
- D_w = Weight for demographic relevance
- P_a = Player age or another demographic measure

Application: Helps tailor loyalty programs that are more effective for specific demographic groups.

8. Comprehensive Multi-Channel Loyalty Score (CMLS)

Combines all channel-specific scores for a unified loyalty metric.

$$CMLS = \frac{\sum_{k=1}^{p}(S_k \times W_k)}{p}$$

Where:

- S_k = Loyalty score for each channel (slots, table games, restaurants, etc.)
- W_k = Weight assigned to each channel for overall contribution to loyalty
- p = Total number of channels

Application: Provides a holistic view of player loyalty across all casino offerings, helping guide reinvestment strategies.

Pros and Cons of Each Model

Model	Pros	Cons
OLS	Easy to calculate; gives a broad overview	May lack depth for individual channels
PSGL	High accuracy for slot player segmentation	Limited to slot game insights
TGEI	Useful for table game strategies	Ignores other types of player engagement
IBL	Captures non-gambling loyalty factors	May require detailed data
RSL	Highlights high-spending dining/shopping customers	Ignores gambling behavior
EPL	Focused on event engagement	Limited to event-based insights
DALS	Customizable for demographics	Requires accurate demographic data
CMLS	Comprehensive view of loyalty	Complex to implement; needs thorough data

By applying a few of these predictive loyalty formulas or variations, casinos can create more tailored and effective marketing campaigns, better allocate marketing spend, and optimize overall player engagement. Each formula focuses on different aspects of player behavior and offers unique insights that can be used for strategic decision-making.

Key Takeaways

1. **Loyalty Programs Drive Engagement and Retention:**
   A well-designed loyalty program encourages players like Emma to return regularly by offering rewards, bonuses, and personalized promotions that keep them engaged.
2. **Personalization is Key to Unlocking Loyalty:**
   By tailoring rewards and experiences to each player's preferences, casinos can create loyalty programs that feel relevant and engaging, increasing player satisfaction and long-term loyalty.
3. **Tiered Loyalty Programs Foster Progression and Achievement:**
   Tiered loyalty programs motivate players to engage more deeply with the casino by offering progressively better rewards as players move through different levels of the program.
4. **Exclusive Experiences Build Emotional Loyalty:**
   Offering exclusive rewards and VIP experiences for high-tier players strengthens emotional loyalty, making players like Emma feel valued and increasing their attachment to the brand.
5. **The Future of Loyalty Programs is Personalized and Immersive:**
   As AI, AR, and predictive analytics continue to evolve, loyalty programs will become even more personalized and immersive, offering dynamic experiences that keep players engaged over the long term.

7.

AI AT PLAY

AI is Transforming Player Interactions!

Emma sits down at her favorite slot machine, unaware of the sophisticated AI systems quietly working behind the scenes. The casino's AI knows her preferences from past visits and recommends a new game she's likely to enjoy based on her playing habits. Later, when she pauses to take a break, a notification pops up on her phone: "Keep the fun going with 50 free spins on your favorite game!" Emma feels valued and engaged, but what she doesn't realize is that AI has been tracking her behavior, analyzing her preferences, and tailoring her entire experience to keep her engaged and satisfied. AI isn't just changing the way casinos operate—it's revolutionizing the way players like Emma interact with the gaming world.

The Role of AI in Modern Casinos

Artificial intelligence (AI) has emerged as one of the most transformative forces in the casino industry, fundamentally changing how players interact with games, promotions, and the overall casino environment. From personalized game recommendations to dynamic promotions and real-time engagement, AI allows casinos to deliver highly tailored experiences to players like Emma, ensuring that every interaction feels unique and engaging.

In today's competitive gaming landscape, understanding and anticipating player behavior is key to building long-term loyalty. AI allows casinos to harness vast amounts of player data, using machine learning algorithms to create individualized experiences that resonate with each player's preferences and habits. For Emma, this means that every recommendation, bonus, and interaction feels custom-made, increasing her engagement and loyalty.

This chapter explores how AI is transforming player interactions in the casino environment, focusing on personalization, real-time engagement, predictive analytics, and the future of AI-driven gaming experiences.

Personalizing the Player Experience with AI

One of the most impactful ways AI is transforming player interactions is through **personalization**. In the past, casinos relied on generalized promotions and rewards that were broadly distributed to all players, regardless of their individual preferences. Today, AI allows casinos to analyze a player's behavior in real time and deliver personalized experiences that make players like Emma feel uniquely understood.

1. Personalized Game Recommendations

For Emma, AI means that every time she steps onto the casino floor, the system knows exactly which games are most likely to appeal to her. Based on her past activity—whether she tends to play high-volatility slots, prefers table games, or enjoys a particular theme—AI can recommend new games or even suggest variations of her favorites that she hasn't yet tried.

For example, if Emma typically plays adventure-themed slots with high volatility, AI can recommend a new slot machine with similar features, creating a seamless and personalized experience. These recommendations ensure that Emma is always engaged with games that match her preferences, keeping her entertained and increasing her likelihood of returning to the casino.

2. Tailored Promotions and Offers

AI also powers **personalized promotions** that align with Emma's gaming habits, spending patterns, and preferences. Rather than receiving generic offers, Emma is sent promotions that are tailored to her individual behavior. If she frequently plays a particular slot machine, she might receive an offer for free spins or a bonus round on that game. If she tends to visit the casino on weekends, she could be offered exclusive weekend bonuses that encourage her to return at her preferred time.

For example, after analyzing Emma's past visits, the casino's AI system might notice that she often spends more during the evening. In response, the system sends her a personalized promotion offering double loyalty points if she plays during her typical evening hours. This level of personalization not only makes Emma feel valued but also encourages her to return at times when she's most likely to engage.

3. Real-Time Adaptation to Player Behavior

AI's ability to analyze data in real time allows casinos to **adapt the player experience on the fly**. For Emma, this means that the casino can respond to her behavior in the moment, offering personalized rewards or recommendations based on her current activity. If Emma has been playing for an extended period without winning, the AI system might offer her a bonus or free spins to keep her engaged. Alternatively, if Emma hits a big win, the AI might suggest a higher-stakes game to capitalize on her excitement.

Real-time adaptation ensures that Emma's experience is dynamic and responsive, making her feel that the casino is actively paying attention to her behavior and adjusting the experience accordingly. This keeps Emma engaged for longer periods and increases her overall satisfaction.

AI-Driven Player Engagement and Retention

AI is also revolutionizing how casinos **engage and retain** players by predicting their behavior and delivering personalized experiences that encourage long-term loyalty. By analyzing patterns in Emma's activity—such as how often she visits, which games she plays, and how much she spends—AI can predict when she might disengage and take proactive steps to keep her involved.

Predictive Analytics for Player Retention

One of AI's most powerful applications is **predictive analytics**, which allows casinos to forecast player behavior based on past activity. For Emma, this means that the casino's AI system can predict when she's likely to visit again, how long she'll play, and even how much she'll spend. If the system detects that Emma's engagement has decreased—for example, if she hasn't visited in a while or has reduced her spending—the AI can trigger a personalized promotion designed to re-engage her.

For instance, if Emma typically visits the casino once a week but hasn't been back for two weeks, the AI might send her a targeted offer to entice her to return, such as a bonus for her favorite game or a limited-time promotion.

These personalized re-engagement strategies help keep Emma loyal to the casino and prevent her from drifting toward competitors.

Real-Time Loyalty Rewards

AI can also drive **real-time loyalty rewards**, offering players like Emma immediate incentives for their actions in the moment. For example, if Emma reaches a milestone—such as a certain number of spins on a slot machine or a specific level in a game—the AI system can automatically reward her with bonus points, free spins, or even exclusive offers. These real-time rewards create a sense of accomplishment and keep Emma motivated to continue playing.

The ability to deliver instant rewards based on real-time behavior makes Emma's experience more engaging and exciting, reinforcing her loyalty to the casino. Rather than waiting for rewards to accumulate over time, Emma can see the benefits of her loyalty in the moment, increasing her overall satisfaction and likelihood of returning.

Dynamic VIP Experiences

AI is also transforming how casinos manage their **VIP programs**, offering dynamic, personalized experiences for high-value players like Emma. By analyzing Emma's engagement and spending patterns, AI can determine whether she qualifies for VIP status and deliver tailored perks accordingly. For instance, Emma might receive exclusive invitations to VIP events, personalized services, or access to special promotions that are reserved for high-tier players.

For Emma, these personalized VIP experiences make her feel valued and appreciated, strengthening her emotional connection to the casino and encouraging her to remain loyal. AI ensures that the VIP experience is not one-size-fits-all but instead tailored to each player's preferences, increasing the likelihood that high-value players will continue to engage with the brand.

The Role of AI in Enhancing Game Design and Experience

AI is not only transforming player interactions outside of gameplay but is also having a profound impact on **game design and the overall gaming experience**. AI-driven games are increasingly offering more immersive, engaging, and personalized experiences, creating new opportunities for player engagement.

AI-Powered Adaptive Games

AI-powered games are designed to **adapt to a player's skill level and preferences**, offering a personalized gaming experience that evolves over time. For Emma, this means that the game can adjust difficulty, rewards, and challenges based on her performance and preferences, ensuring that the experience remains engaging and tailored to her abilities.

For example, if Emma is playing a slot machine and consistently wins at lower stakes, the game might automatically suggest higher-stakes options or introduce new features to keep her challenged and engaged. Conversely, if Emma is struggling to win, the game might adjust its difficulty level or offer bonus rounds to keep her motivated. This adaptive gameplay creates a dynamic experience that keeps Emma engaged over time, increasing her likelihood of returning to play.

Personalized Game Content

AI also enables **personalized game content**, where the game's themes, characters, and storylines are tailored to each player's preferences. For Emma, this means that the games she plays can be customized to reflect her favorite themes, aesthetics, or gameplay styles, making the experience more immersive and enjoyable.

For instance, if Emma enjoys adventure-themed slots, AI could generate custom game content that includes her favorite themes, characters, or storylines. This level of personalization makes the game feel uniquely tailored to Emma, increasing her engagement and satisfaction with the overall gaming experience.

Enhanced Social and Multiplayer Features

AI is also enhancing the **social and multiplayer aspects of gaming**, allowing players like Emma to connect with others in real time and participate in dynamic, interactive gaming experiences. AI-driven chatbots, virtual hosts, and multiplayer matchmaking systems ensure that players can engage with others in a seamless and immersive environment.

For example, AI-powered virtual hosts can offer real-time support, answer questions, and provide personalized recommendations based on a player's preferences. Additionally, AI-driven matchmaking systems can pair Emma with players of similar skill levels, creating a more enjoyable and competitive multiplayer experience. These social features make Emma feel more connected to the gaming community, increasing her overall engagement and loyalty.

The Future of AI in Player Interactions

As AI technology continues to evolve, the future of AI-driven player interactions will become even more personalized, immersive, and dynamic. Emerging technologies such as **natural language processing (NLP)**, **virtual reality (VR)**, and **augmented reality (AR)** will further enhance the ability of casinos to deliver highly engaging and tailored experiences to players like Emma.

AI-Powered Virtual Assistants and Chatbots

Soon, AI-powered **virtual assistants** and chatbots will play an even larger role in enhancing player interactions. For Emma, this means that she'll be able to ask a virtual assistant for game recommendations, get real-time updates on her loyalty status, or even receive personalized tips on how to improve her gameplay. These AI-driven assistants will provide seamless support and guidance, making Emma's casino experience more intuitive and enjoyable.

Augmented Reality (AR) and Virtual Reality (VR) Integration

AI will also integrate with **AR and VR** technologies to create more immersive gaming experiences. For Emma, this could mean participating in AR-based treasure hunts within the casino or experiencing fully immersive VR slot machines that transport her to a virtual casino environment. These technologies, powered by AI, will blur the lines between physical and digital gaming, creating new opportunities for player engagement.

Hyper-Personalization through AI

As AI continues to advance, the level of **personalization** in player interactions will become even more granular. For Emma, this means that every aspect of her casino experience—from the games she plays to the rewards she receives—will be hyper-personalized to match her unique preferences and behavior. AI will analyze vast amounts of player data to deliver tailored experiences that feel custom-made for each individual, ensuring that Emma's engagement remains high over the long term.

Key Takeaways

1. **AI is Revolutionizing Personalization:**
   AI allows casinos to deliver highly personalized experiences, from game recommendations to tailored promotions, ensuring that players like Emma feel valued and engaged at every interaction.
2. **Predictive Analytics Drives Player Retention:**
   By predicting player behavior, AI helps casinos retain players like Emma by delivering timely, personalized offers that encourage repeat visits and long-term loyalty.
3. **Real-Time Rewards Enhance Player Engagement:**
   AI-powered real-time rewards keep players motivated and engaged by offering instant incentives based on their current actions, creating a dynamic and exciting experience.
4. **AI is Transforming Game Design:**
   AI-driven games offer adaptive gameplay, personalized content, and enhanced social features, ensuring that players like Emma enjoy a more immersive and tailored gaming experience.
5. **The Future of AI in Casinos is Immersive and Hyper-Personalized:**
   As AI integrates with AR, VR, and other emerging technologies, the future of player interactions will become even more personalized, immersive, and engaging.

8.

THE FUTURE ARRIVES

Tech Reshaping Gaming

Emma walks through the casino, but it doesn't feel like the same place she visited just a few years ago. The slot machines have evolved into interactive experiences with virtual reality (VR) features, the digital displays are seamlessly integrated with augmented reality (AR), and even the loyalty program is smarter, with personalized rewards delivered in real-time. As she engages with a new table game, her AI-powered virtual assistant provides helpful tips on strategy and offers her a personalized bonus based on her past play. This is no longer just a casino—it's an immersive, tech-driven entertainment environment where cutting-edge innovations create experiences Emma couldn't have imagined before.

Emerging Technologies

The future of gaming is here, and it's defined by a wave of **emerging technologies** that are transforming every aspect of the casino environment—from the way games are played to how players like Emma interact with the casino itself. As these technologies advance, they are creating more personalized, engaging, and immersive experiences, blurring the lines between physical and digital worlds. The traditional casino floor is being reimagined, leveraging augmented reality (AR), virtual reality (VR), artificial intelligence (AI), advanced data analytics, and other innovations to enhance player experiences, increase engagement, and drive loyalty.

Let's explore some of the most impactful emerging technologies that are reshaping the gaming industry. From immersive VR games to AI-driven player interactions, these technologies are not just enhancing the casino experience—they are revolutionizing it.

Virtual Reality (VR)

One of the most exciting technologies transforming the gaming experience is **virtual reality (VR)**. VR technology allows casinos to offer fully immersive gaming environments that transport players into virtual worlds, where they can engage with games in a whole new way. For Emma, this means stepping into a VR casino where she can interact with slot machines, table games, and even other players in a virtual environment that feels as real as the physical world.

VR Slot Machines and Table Games

Imagine Emma putting on a VR headset and entering a fully immersive, 3D casino where she can walk around, choose games, and interact with the environment. She might sit at a virtual blackjack table where the dealer is an AI-driven avatar, or explore a VR slot machine that transports her to a fantasy world, where each spin unlocks new visual and interactive elements.

The ability to create immersive environments through VR transforms gaming into a more engaging, multisensory experience. Instead of simply watching reels spin on a screen, Emma is now part of the action, surrounded by rich visuals, sound effects, and interactive elements that make the game more exciting and memorable. For casinos, VR gaming offers a unique way to engage players and create experiences that are difficult to replicate in traditional gaming settings.

VR Social Casino Experiences

VR also opens up new opportunities for **social interactions** in the gaming environment. With VR technology, Emma can meet and play with other players in real time, regardless of their physical location. This creates a more social, interactive casino experience where players can engage with one another in a virtual world, participate in multiplayer games, and even attend virtual events.

For example, Emma might join a virtual poker tournament where she plays against real people from around the world, all of whom are represented by avatars in the virtual casino. This social element adds a new layer of engagement, making the casino experience more interactive and dynamic.

Augmented Reality (AR)

AR is another transformative technology that is reshaping the gaming landscape by blending the physical world with digital elements. Unlike VR, which creates entirely virtual environments, AR overlays digital information and graphics onto the real world, enhancing the physical casino environment with interactive elements.

AR-Enhanced Slot Machines and Games

For Emma, AR can make the physical casino floor come alive. Imagine playing a slot machine that uses AR to project interactive graphics and animations into the space around the machine. As she spins the reels, AR elements might appear, creating a more immersive experience where game characters come to life, and bonus rounds are presented in a dynamic, interactive way.

AR can also enhance table games by adding digital layers of information to the physical game environment. For example, when Emma sits down at a blackjack table, AR could project a digital overlay that shows the game's odds, her historical performance, and personalized tips based on her playing style. These AR enhancements make the game more interactive and engaging, offering players real-time information that can improve their gameplay experience.

AR Treasure Hunts and Promotions

AR is also being used to create unique **promotional experiences** that blend the digital and physical worlds. For example, Emma might participate in an AR-powered treasure hunt, where she uses her smartphone to search for hidden digital rewards scattered throughout the casino. By following clues and scanning QR codes, Emma can unlock bonuses, free play, or other prizes, making the promotion an interactive, gamified experience that encourages her to explore different areas of the casino.

These AR-driven promotions add a sense of adventure and fun to the casino experience, encouraging players like Emma to engage with the casino environment in new and exciting ways.

Artificial Intelligence (AI)

AI has already begun to transform player interactions in the gaming world, and its impact will only grow in the future. AI technology allows casinos to analyze vast amounts of player data, delivering personalized experiences that cater to each player's preferences, habits, and behaviors. For Emma, this means that every aspect of her experience at the casino—from game recommendations to promotions—is tailored specifically to her.

AI-Powered Personalized Recommendations

AI systems use machine learning algorithms to track Emma's gameplay, preferences, and spending patterns, allowing the casino to offer personalized game recommendations that align with her tastes. For example, if Emma regularly plays high-volatility slots with fantasy themes, the AI might recommend a new slot machine with similar features, ensuring that every recommendation feels relevant and engaging.

These personalized recommendations keep Emma engaged with the casino, as she feels that the system understands her preferences and is offering games that match her interests. The result is a more tailored, enjoyable experience that encourages long-term loyalty.

AI-Driven Customer Service

AI is also transforming customer service in the casino environment through the use of **AI-powered virtual assistants** and chatbots. For Emma, this means that she can receive instant answers to her questions, whether she's looking for information on promotions, checking her loyalty status, or seeking game recommendations.

For example, Emma could ask a virtual assistant for tips on how to improve her strategy at the blackjack table, and the AI system would provide personalized suggestions based on her gameplay history. These AI-driven interactions make the casino experience more seamless and accessible, as Emma can receive real-time support without having to wait for human intervention.

Advanced-Data Analytics

Data analytics plays a crucial role in the modern casino, allowing operators to collect and analyze data on player behavior, preferences, and engagement. This data-driven approach enables casinos to make informed decisions about game offerings, promotions, and player experiences, ultimately improving the overall gaming ecosystem.

Predictive Analytics for Player Behavior

By leveraging **predictive analytics**, casinos can forecast Emma's behavior based on her past activity, allowing them to deliver timely promotions and incentives that keep her engaged. For instance, if the data shows that Emma typically visits the casino on Friday evenings, the casino might send her a personalized offer on Friday afternoon, encouraging her to return that night.

Predictive analytics also help casinos identify when players like Emma are at risk of disengaging. If the system detects that Emma hasn't visited in a while or has reduced her spending, the casino can proactively offer her a promotion designed to re-engage her, such as a bonus for her favorite game. This data-driven approach helps retain players by addressing potential churn before it occurs.

Optimizing Game Offerings and Promotions

Advanced data analytics also allow casinos to **optimize their game offerings and promotions** based on player preferences. For example, the casino might analyze gameplay data to determine which slot machines are most popular among certain demographics or which promotions drive the highest engagement. This information allows the casino to adjust its offerings to better align with player preferences, ensuring that the games and promotions available are those that resonate most with players like Emma.

For Emma, this means that the casino's game selection is constantly evolving based on player demand, ensuring that there are always fresh, exciting options available that match her interests.

Contactless and Touch-Free Technologies

As casinos adapt to the changing landscape of the hospitality industry, **contactless technologies** are becoming more prevalent, offering players a safer and more convenient way to interact with games, payments, and services. For Emma, contactless technology enhances her experience by reducing friction and improving the ease of gameplay.

Mobile Payments and Digital Wallets

Contactless payments, such as **mobile payments and digital wallets,** allow Emma to make transactions quickly and securely without the need for cash or physical cards. She can use her smartphone to deposit funds, purchase chips, or redeem loyalty points, all with a few taps on her screen. This seamless payment experience not only improves convenience but also enhances safety by minimizing physical contact.

For Emma, the ability to manage her funds digitally means she can spend more time enjoying the casino experience and less time dealing with traditional payment methods.

Touch-Free Game Controls

In addition to contactless payments, **touch-free game controls** are becoming more common in casinos, allowing players like Emma to interact with games using gestures or voice commands rather than physical buttons. For example, Emma could use hand gestures to spin the reels on a slot machine or use voice commands to place bets at a digital roulette table.

These touch-free controls offer a more hygienic, intuitive way to engage with games, especially in a post-pandemic world where safety and cleanliness are top priorities.

As emerging technologies continue to evolve, the future of gaming will be defined by **increasingly immersive, personalized, and interactive experiences**. For players like Emma, this means that every aspect of the casino environment—from game design to promotions—will be tailored to her preferences, creating a more engaging and enjoyable experience. As AI technology advances, casinos will be able to deliver even more **real-time personalized experiences**, where every interaction Emma has with the casino is tailored to her current behavior. Whether it's real-time promotions, dynamic game recommendations, or instant rewards, AI will make Emma's experience feel more responsive and customized than ever before.

Key Takeaways

1. **Virtual Reality (VR) is Creating Immersive Gaming Environments:**
   VR allows players like Emma to step into fully immersive, 3D casino environments where they can engage with games in a

more interactive and dynamic way, transforming the traditional gaming experience.

2. **Augmented Reality (AR) Enhances the Physical Casino Floor:**
AR technology overlays digital elements onto the real world, making games more interactive and engaging, while offering unique promotional experiences like AR-driven treasure hunts.

3. **AI is Powering Personalization and Real-Time Engagement:**
AI allows casinos to analyze player data and deliver personalized game recommendations, promotions, and rewards in real time, ensuring that every experience is tailored to players like Emma.

4. **Advanced Data Analytics is Driving Informed Decision-Making:**
Data analytics helps casinos optimize game offerings, promotions, and player experiences based on player preferences and behavior, creating a more engaging and rewarding gaming environment.

5. **Contactless Technologies Enhance Safety and Convenience:**
Contactless payments, mobile wallets, and touch-free game controls offer players like Emma a safer, more convenient way to interact with the casino, improving the overall experience.

9.

DATA THAT MATTERS

Using Analytics to Drive Success

Emma enters the casino, unaware that her every move is being recorded and analyzed—her favorite games, the frequency of her visits, her spending patterns, and even how long she plays each game. As she engages with her favorite slot machine, the casino's data systems are processing this information in real time, delivering insights that will shape future offers and promotions tailored just for her. When she receives a notification offering extra loyalty points for playing a new game that aligns with her preferences, she smiles. Little does she know that behind the scenes, data analytics is driving the entire experience, ensuring that every interaction feels personal and engaging.

The Role of Data Analytics

In today's competitive gaming landscape, **data is one of the most valuable assets** casinos have at their disposal. By leveraging data analytics, casinos can gain deep insights into player behavior, preferences, and trends, allowing them to make more informed decisions that enhance player engagement, improve operational efficiency, and drive long-term success. For players like Emma, this means that their casino experience feels more personalized, with promotions, game recommendations, and rewards tailored to their unique preferences.

Let's explore how data analytics transforms the casino industry, from optimizing player engagement and retention to enhancing operational performance and improving marketing strategies. By using data that truly matters, casinos can make smarter decisions that lead to greater player satisfaction, increased revenue, and long-term growth.

The Importance of Data

Data analytics is no longer just a luxury for casinos—it's a **strategic necessity**. Every interaction that Emma has with the casino generates valuable data, from the games she plays and the promotions she engages with, to the frequency of her visits and the length of her play sessions. This data provides casinos with a goldmine of insights into what drives player behavior, what keeps players coming back, and how to optimize the overall gaming experience.

By collecting, analyzing, and acting on this data, casinos can:

- **Understand player preferences:** What games do players like Emma prefer? How much do they typically spend, and what promotions resonate with them?
- **Improve player retention:** How can casinos encourage Emma to visit more frequently, play longer, and engage with new games?
- **Optimize marketing efforts:** What marketing campaigns are most effective at driving player engagement? How can promotions be better targeted to appeal to individual players?

- **Enhance operational efficiency:** How can staffing, game placement, and floor layout be optimized to improve player satisfaction and maximize revenue?

Data analytics allows casinos to answer these questions and more, creating a more tailored, efficient, and profitable gaming environment.

Player Behavior Through Data

One of the most powerful applications of data analytics in casinos is its ability to **analyze player behavior** in real-time. By collecting data on how players like Emma interact with games, promotions, and the casino environment, casinos can better understand player preferences and habits, allowing them to create more engaging experiences.

Tracking Player Preferences

For Emma, this means that every game she plays, every dollar she spends, and every promotion she engages with is recorded and analyzed. By tracking this data, the casino can build a detailed profile of Emma's preferences, allowing it to deliver personalized experiences that resonate with her. For example, if Emma consistently plays certain types of slot machines with adventure themes and high volatility, the casino's data systems will take note, ensuring that future game recommendations and promotions align with these preferences.

This level of personalization not only makes Emma's experience more enjoyable but also increases her likelihood of staying engaged with the casino over the long term. When players feel that the casino understands their preferences, they are more likely to return and participate in future promotions, driving both loyalty and revenue.

Predicting Player Behavior

Data analytics also allows casinos to **predict player behavior** based on past interactions. By analyzing historical data, the casino can forecast when Emma is most likely to visit, how long she will play, and what games she will engage with. This predictive capability enables the casino to tailor promotions and offers to Emma at the right time, encouraging her to visit more frequently and stay longer.

For example, suppose Emma typically visits on weekends and plays for several hours. In that case, the casino might send her a promotion offering a bonus for visiting on a Saturday evening, timed to coincide with her usual play schedule. This type of data-driven promotion increases the likelihood that Emma will respond positively, driving higher engagement and retention.

Enhancing Player Retention and Engagement

Data analytics is also a powerful tool for **improving player retention and engagement**. Casinos can develop strategies that encourage repeat visits, longer play sessions, and deeper interactions with the casino environment by understanding what motivates players like Emma to stay engaged.

Identifying At-Risk Players

One of the key challenges for casinos is identifying when players are at risk of disengaging. With data analytics, casinos can **detect declining engagement patterns**—such as fewer visits, shorter play sessions, or reduced spending—and take proactive steps to re-engage those players. Emma might receive a personalized offer if the system detects that she hasn't visited the casino in a while or if her engagement levels have decreased.

For instance, if Emma typically visits once a week but hasn't returned for two weeks, the casino's data analytics system might flag her as an at-risk player. In response, the system could send her a personalized promotion offering a bonus or free play for her next visit, encouraging her to return and re-engage with the casino.

Delivering Real-Time Rewards and Incentives

Data analytics also enables casinos to **deliver real-time rewards and incentives** based on a player's current behavior. For Emma, this means that her actions—whether it's playing a certain number of spins on a slot machine or reaching a milestone in a table game—can trigger immediate rewards. These real-time incentives create a sense of instant gratification, keeping Emma engaged and motivated to continue playing.

For example, if Emma has been playing a particular slot machine for an extended period, the data system might automatically trigger a reward—such as free spins or a loyalty point multiplier—to keep her engaged and excited. These timely rewards increase player satisfaction and drive longer play sessions and higher spending.

Optimizing Marketing and Promotions

Data analytics is also transforming the way casinos approach **marketing and promotions**. Rather than relying on broad, one-size-fits-all campaigns, casinos can use data to create **targeted, personalized marketing strategies** that resonate with individual players like Emma.

Personalizing Promotions

By analyzing player data, casinos can **personalize promotions** to match each player's preferences and behaviors. For Emma, this means receiving promotions that are specifically tailored to her favorite games, play patterns, and spending habits. If Emma regularly plays a specific type of slot machine, she might receive a promotion offering free spins or a bonus on that game. If she tends to visit the casino at certain times of the week, the promotion might be timed to coincide with her usual visits.

These personalized promotions increase the likelihood that Emma will engage with the offer, as they feel relevant and aligned with her interests. This data-driven approach to marketing ensures that promotions are not just generic offers, but carefully crafted incentives that appeal to each player's unique preferences.

Optimizing Campaign Effectiveness

Data analytics also allows casinos to **track the effectiveness of marketing campaigns** in real time, providing valuable insights into which promotions are driving the most engagement. For example, the casino can analyze how many players, like Emma, respond to a specific promotion, how much they spend after engaging with the offer, and whether the promotion leads to longer play sessions or repeat visits.

By continuously analyzing campaign performance, casinos can refine their marketing strategies to focus on the promotions that deliver the highest returns. For Emma, this means receiving more of the promotions she enjoys and fewer of the offers that don't resonate with her, creating a more personalized and engaging experience.

Improving Operational Efficiency

Data analytics isn't just about improving player engagement but also about enhancing the casino's operational efficiency. Casinos can optimize their operations by analyzing data on game performance, staffing levels, and player traffic patterns to create a more seamless and enjoyable experience for players like Emma.

Optimizing Game Placement and Floor Layout

One of the ways data analytics can improve operational efficiency is by **optimizing the placement of games and the layout of the casino floor**. By analyzing data on which games are most popular, how players move through the casino, and where they spend the most time, casinos can adjust their floor layout to maximize player engagement and revenue.

For example, if data shows that Emma and other players tend to spend more time in certain areas of the casino, the casino can place high-performing games in those areas to increase engagement. Similarly, if certain games are underperforming, the casino might move them to a different location or replace them with more popular options.

Enhancing Staffing and Service Levels

Data analytics also allows casinos to **optimize staffing levels** based on player traffic patterns and peak times. By analyzing when players like Emma are most likely to visit and how long they stay, the casino can ensure that it has the right number of staff on hand to provide excellent service. This data-driven approach to staffing ensures that players receive prompt, personalized service, improving overall satisfaction.

For Emma, this means that her experience at the casino feels seamless and efficient. Whether she's ordering a drink, interacting with a dealer, or receiving customer support, the casino is able to provide the right level of service at the right time, enhancing her overall experience.

Leveraging Data for Long-Term Success

The true value of data analytics lies in its ability to drive **long-term success** for casinos by continuously improving player engagement, operational efficiency, and overall profitability. By leveraging data to make informed decisions, casinos can stay ahead of the competition and create a more dynamic, engaging environment for players like Emma.

Building a Data-Driven Culture

For casinos, building a **data-driven culture** is essential for unlocking the full potential of data analytics. This means not only collecting and analyzing data but also using insights to inform decision-making at every level of the organization. By integrating data analytics into everything from marketing strategies to floor layout design, casinos can create a more responsive and efficient gaming environment that adapts to player needs in real time.

For Emma, this data-driven approach translates into a more personalized and enjoyable experience. From the moment she steps onto the casino floor to the time she leaves, every interaction feels tailored to her preferences, making her more likely to return and engage with the casino in the future.

Driving Innovation and Growth

As casinos continue to adopt and refine their data analytics capabilities, they can use these insights to drive **innovation**

and growth. Whether it's introducing new games that resonate with players, developing more effective marketing campaigns, or optimizing operational performance, data analytics provides the foundation for continuous improvement and long-term success.

For players like Emma, this means that the casino experience will continue to evolve, offering new, exciting opportunities for engagement and entertainment.

Key Takeaways

1. **Data Analytics Provides Deep Insights into Player Behavior:**
   By tracking and analyzing player behavior, preferences, and engagement, casinos can deliver personalized experiences that keep players like Emma engaged and loyal.
2. **Real-Time Data Drives Immediate Rewards and Engagement:**
   Data analytics allows casinos to offer real-time rewards and incentives based on a player's current activity, creating a more dynamic and exciting experience.
3. **Personalized Marketing Increases Campaign Effectiveness:**
   Data-driven marketing strategies ensure that promotions are tailored to each player's preferences, increasing the likelihood of engagement and driving long-term loyalty.
4. **Operational Efficiency is Improved with Data Analytics:**
   By analyzing data on game performance, traffic patterns, and staffing levels, casinos can optimize their operations to create a more seamless and enjoyable experience for players.
5. **Data Analytics is Key to Long-Term Success and Growth:**
   Leveraging data to make informed decisions allows casinos to continuously improve player engagement, operational performance, and overall profitability, ensuring long-term success.

10.

DATA'S DILEMMA

The Ethical Challenges of Using Player Data

Emma strolls through the casino, unaware that every tap on a slot machine, every swipe of her loyalty card, and every click on the mobile app is being tracked, stored, and analyzed. Behind the scenes, her data is being used to personalize her experience, sending her tailored promotions and recommending games that align with her preferences. Emma enjoys the personalized offers and the seamless experience, but there's a tradeoff: her data is continuously collected and used by the casino. This creates a significant question—how is her data being used, and is it being protected? In today's data-driven world, the ethical use of player data is no longer just a business consideration—it's a moral responsibility.

The Power and Responsibility

In modern casinos, **data is a powerful tool** that enables personalized experiences, targeted marketing, and optimized operations. However, with this power comes the ethical responsibility of ensuring that player data—like Emma's—is collected, stored, and used in ways that are transparent, secure, and respectful of privacy. The widespread use of data analytics, AI, and predictive technologies has created unprecedented opportunities for casinos, but it has also raised critical ethical questions about how player data is handled.

For Emma and millions of other players, the collection of personal data is a double-edged sword. On the one hand, it leads to a more enjoyable, tailored experience that aligns with her gaming preferences. On the other hand, it poses potential risks if that data is misused or inadequately protected. As casinos increasingly rely on data to drive engagement, they must navigate the ethical dilemmas that arise from the use of player information.

This chapter explores the **ethical challenges** associated with using player data in the casino industry. We'll dive into issues such as privacy, consent, transparency, data security, and fairness and examine how casinos can balance the benefits of data-driven personalization with their ethical obligations to players like Emma.

Privacy

At the heart of the ethical dilemma surrounding player data is the issue of **privacy**. As casinos collect more information about their players, they must ensure that they are not infringing on individuals' right to privacy. For Emma, this means that while she enjoys the personalized experiences that data-driven technologies offer, she also expects her personal information to be treated with care and respect.

The Scope of Data Collection

Modern casinos collect vast amounts of data about their players, ranging from **personal details** (such as age, gender, and contact information) to **behavioral data** (such as game preferences, spending patterns, and visit frequency). For Emma, this means that every interaction she has with the casino—whether in person or online—contributes to her digital profile.

While this data is invaluable for creating personalized experiences, it also raises concerns about how much information is being collected and whether all of it is necessary. Do casinos really need to track every aspect of Emma's behavior, or is some data collection excessive? The ethical challenge lies in striking the right balance between gathering enough data to enhance the player experience without overstepping privacy boundaries.

Privacy by Design

To address privacy concerns, casinos must adopt a **"privacy by design"** approach, where player privacy is embedded into every aspect of data collection and use. This means that casinos should only collect the data they truly need, limit how long that data is stored, and ensure that it is used solely for the purpose of enhancing the player experience. For Emma, this would provide reassurance that her data is not being collected unnecessarily or used in ways that she hasn't consented to.

Implementing privacy by design also requires robust data protection measures, such as anonymizing player data to reduce the risk of it being linked to specific individuals. This ensures that even if data is collected and analyzed, Emma's privacy remains intact.

Consent

Another key ethical issue in the use of player data is **consent**—the idea that players like Emma should have control over what data is collected about them and how it is used. In a data-driven casino environment, players must be fully informed about how their data will be used and are given the opportunity to provide or withhold consent.

Informed Consent

For Emma, informed consent means that before the casino collects any of her personal information, she is made aware of what data will be collected, how it will be used, and for what purposes. This information should be communicated in a clear, straightforward manner, without the use of legal jargon or hidden clauses. Too often, consent is obtained through lengthy, complex terms and conditions that players may not fully understand. To ensure ethical data use, casinos must make it easy for players like Emma to understand what they are agreeing to.

In addition, players should have the ability to **opt out** of certain types of data collection or marketing practices. For instance, if Emma prefers not to receive personalized promotions or wants to limit the amount of data the casino collects about her, she should be able to adjust her preferences accordingly. Providing players with control over their data is essential for maintaining trust and ensuring that consent is meaningful rather than just a formality.

Consent and Personalization Trade-offs

There is often a trade-off between personalization and privacy, as casinos need player data to deliver tailored experiences. For Emma, this might mean that if she chooses to limit the data collected about her, she may not receive the same level of personalization in her casino experience. The ethical challenge is ensuring that players are aware of this trade-off and can make informed decisions about how much data they are willing to share in exchange for personalized experiences.

Transparency

Transparency is a cornerstone of ethical data use. Casinos must be open about how they collect, store, and use player data, ensuring that players like Emma have a clear understanding of what happens to their information once it's collected. Without transparency, there's a risk that players may feel their data is being exploited or misused, which can erode trust and damage the casino's reputation.

Clear Communication

For Emma, transparency means that the casino provides clear, accessible information about its data practices, including how her data will be used to personalize her experience and what safeguards are in place to protect her information. This information should be easy to find, whether through the casino's website, mobile app, or in-person interactions. Transparency also extends to explaining what types of data are being collected and for what specific purposes—whether it's for improving the gaming experience, offering personalized promotions, or optimizing casino operations.

Data Sharing and Third Parties

One of the most critical transparency issues is **data sharing**—specifically, whether player data is being shared with third-party companies. For Emma, it's important to know if her data is being sold or shared with advertisers, analytics firms, or other external entities. If the casino partners with third-party companies for marketing or data analysis, Emma should be informed about who has access to her data and for what purpose.

Being transparent about data sharing practices helps build trust with players and ensures that casinos remain accountable for how player data is handled.

Data Security

As casinos collect more data, they also face increased responsibility to protect that information from breaches, theft, or misuse. **Data security** is an ethical imperative, as players like Emma trust that their personal and financial information is being stored securely and that it won't fall into the wrong hands.

Protecting Sensitive Data

For Emma, her personal information—such as her name, contact details, and financial transactions—is highly sensitive. If this data were to be compromised, it could result in identity theft, fraud, or other harmful consequences. As a result, casinos must implement strong data security measures, including encryption, secure data storage, and regular security audits, to ensure that player data is protected from cyberattacks and breaches.

Security is particularly important for casinos, which handle not only personal data but also **financial transactions**. Ensuring the security of payment systems, loyalty programs, and online gaming platforms is essential to maintaining player trust.

Data Breach Response

In the event of a data breach, casinos have an ethical responsibility to **respond quickly and transparently**. For Emma, this means that if her data is compromised, the casino should notify her immediately, provide information about what data was affected, and offer guidance on how to protect herself from further harm. This proactive approach to breach management is essential for maintaining player trust and minimizing the impact of security incidents.

Fairness

As casinos rely on data analytics and AI to personalize experiences and make decisions, they must be vigilant about ensuring **fairness** in how data is used. One of the ethical challenges of data-driven systems is the potential for **bias and discrimination**, especially if the algorithms that drive personalization are not carefully monitored and designed.

Avoiding Discriminatory Practices

For Emma, fairness means that the casino's data-driven systems treat her equitably, without bias based on factors such as age, gender, or socioeconomic status. For example, AI algorithms that recommend games or promotions should be designed to avoid **discriminatory patterns**, ensuring that all players receive equal opportunities and access to rewards.

Bias can inadvertently be introduced into data-driven systems if the algorithms rely on incomplete or biased data sets. To address this, casinos must ensure that their AI systems are regularly audited and tested for fairness and that any biased outcomes are corrected.

Ethical Use of Predictive Analytics

Predictive analytics, while powerful, can also raise ethical questions if used to manipulate player behavior in harmful ways. For example, using predictive analytics to encourage players like Emma to spend more than they can afford could be seen as exploitative. To maintain ethical standards, casinos must use data responsibly, ensuring that predictive analytics are applied in ways that enhance the player experience without taking advantage of vulnerabilities or encouraging unhealthy gambling behaviors.

Navigating Data's Ethical Dilemma

As casinos continue to embrace data-driven technologies, they must navigate a range of ethical dilemmas related to the use of player data. For Emma, the key issues are privacy, consent, transparency, security, and fairness. By addressing these challenges head-on, casinos can build trust with their players and create data practices that are both ethical and effective.

Building Trust Through Ethical Data Practices

To build trust, casinos must prioritize transparency and ensure that players understand how their data is being used. This involves clear communication, informed consent, and the opportunity for players to control their own data. By taking a proactive approach to data ethics, casinos can create a more trusting and loyal player base.

Protecting Player Data

Data security is paramount, and casinos must invest in the tools and technologies needed to protect sensitive player information. This includes regular security audits, encryption, and robust breach response protocols. For Emma, knowing that her data is secure will give her peace of mind and encourage her to continue engaging with the casino.

Key Takeaways

1. **Privacy is Key to Ethical Data Use:**
   Casinos must balance data collection with respect for player

privacy, ensuring that only necessary data is gathered and used responsibly.
2. **Informed Consent Empowers Players:**
Players like Emma should have control over their data, with the ability to provide or withhold consent for specific data collection and usage practices.
3. **Transparency Builds Trust:**
Casinos must be transparent about how player data is collected, used, and shared, providing clear information about data practices and third-party partnerships.
4. **Data Security is Essential for Protecting Players:**
Strong data security measures are critical to safeguarding sensitive player information and ensuring that players feel confident in engaging with the casino.
5. **Fairness in Data Use Prevents Bias and Discrimination:**
Data-driven systems must be designed to avoid biased outcomes and ensure that all players are treated fairly and equitably in the gaming environment.

11.

PRIVACY AT STAKE

Ensuring Privacy with Compliance

Emma enjoys the casino's seamless experience—personalized game recommendations, timely promotions, and a loyalty program that seems custom-tailored just for her. But what Emma doesn't see is the vast amount of personal data that's being collected, analyzed, and stored behind the scenes. While this data makes her experience more engaging, it also brings up important questions: how is her privacy being protected, and what steps are the casino taking to ensure compliance with the latest privacy laws? In the age of data-driven gaming, where every move a player makes is captured and stored, safeguarding privacy has become a critical responsibility for casinos.

The Growing Importance of Privacy

As casinos increasingly rely on data to enhance the player experience, **privacy** has become one of the most significant concerns in the industry. Players like Emma are rightfully concerned about how their personal information is being used, whether it's being protected and whether they have control over it. The rise of global data protection regulations like the **General Data Protection Regulation**

(GDPR) in Europe and the **California Consumer Privacy Act (CCPA)** in the United States has made it clear that businesses—including casinos—must take privacy seriously. Non-compliance can lead to hefty fines, legal consequences, and most importantly, a loss of player trust.

Let's explore the key principles and challenges related to ensuring privacy in the casino industry, including data protection, compliance with regulations, transparency with players, and the implementation of privacy-enhancing technologies. It also highlights best practices for casinos to ensure they are compliant with laws and respected by players who value their privacy.

Data Privacy

For casinos, **protecting player privacy** is both a legal and ethical responsibility. Every piece of personal information collected—whether it's Emma's contact details, payment information, or game preferences—must be handled with care. Privacy regulations such as the GDPR and CCPA set clear guidelines for how businesses should collect, store, and use personal data. Failure to comply with these regulations can result in significant fines and damage to the casino's reputation.

Understanding Key Privacy Regulations

GDPR:
The **General Data Protection Regulation** (GDPR) is one of the strictest privacy laws in the world and applies to any business that handles the personal data of European citizens, regardless of where the business is located. For casinos, this means that if Emma is an EU citizen, her data must be handled in accordance with GDPR guidelines. Some of the key requirements of GDPR include:

- **Data Minimization:** Collect only the data necessary for a specific purpose.
- **Right to Access and Erasure:** Emma has the right to access the data the casino holds on her and can request its deletion if it's no longer necessary for the intended purpose.
- **Data Breach Notification:** In the event of a data breach, the casino must notify Emma and the relevant authorities within 72 hours.

CCPA:

The **California Consumer Privacy Act** (CCPA) grants residents of California, like Emma if she were a Californian player, similar rights over their personal data. Under the CCPA, Emma has the right to:

- Know what personal information is being collected about her.
- Request the deletion of her personal information.
- Opt-out of the sale of her personal data to third parties.
- Receive equal service even if she exercises her privacy rights (no discrimination for opting out).

Casinos must ensure they are fully compliant with these regulations to avoid penalties and protect players like Emma from potential misuse of their data.

Ethical Considerations

Beyond legal requirements, there's an ethical dimension to **data privacy**. Players expect casinos to handle their data responsibly, even if there's no explicit legal requirement. For Emma, trust is paramount—she expects the casino to safeguard her personal information, not share it without her consent, and use it only for purposes she agrees to. Ethical data handling can enhance player loyalty, while unethical practices can quickly erode trust and lead to player churn.

Transparency

One of the most critical elements of ensuring privacy is **transparency**. Players like Emma should always know how their data is being collected, used, and stored. Transparency builds trust, making players more comfortable sharing their information with the casino because they understand how it will be used and what protections are in place.

Clear Privacy Policies

For Emma, transparency starts with a **clear and concise privacy policy**. The privacy policy should be easily accessible on the casino's website or mobile app, written in plain language, and free of legal jargon. It should explain:

- What types of data are being collected (e.g., personal information, gameplay data, financial information).
- How the data will be used (e.g., for personalized promotions, loyalty programs, or fraud detection).
- Whether the data will be shared with third parties and for what purpose.
- Emma's rights regarding her data, such as the right to access, delete, or correct her personal information.
- How long her data will be stored and when it will be deleted.

Providing players like Emma with this information upfront empowers them to make informed decisions about their data, fostering a sense of control and trust.

Communicating Changes in Data Policies

Casinos must also be proactive in communicating any **changes to data policies**. If the casino decides to start using player data in new ways—such as introducing a new analytics platform or partnering with third-party marketing firms—Emma should be notified of these changes and given the opportunity to adjust her privacy settings. Failing to communicate these changes can lead to mistrust and could even result in legal repercussions if it violates privacy regulations.

Data Security

It is at the core of privacy protection. No matter how robust a privacy policy is, it's meaningless if the casino cannot protect player data from breaches, theft, or unauthorized access. For players like Emma, data security ensures that her personal information—such as her credit card details, address, and gaming history—is protected from cyberattacks and data breaches.

Implementing Strong Data Security Measures

To protect player data, casinos must implement **strong security protocols** that safeguard sensitive information. These include:

- **Encryption:** Personal and financial data should be encrypted both in transit (while being transmitted) and at rest (when stored on servers). For Emma, this means that her credit card information is protected from unauthorized access when she makes a payment or registers for an online account.
- **Multi-Factor Authentication (MFA):** Adding an extra layer of security, such as requiring Emma to enter a one-time code sent to her phone in addition

to her password, can significantly reduce the risk of unauthorized access.
- **Regular Security Audits:** Casinos should conduct regular security audits to identify vulnerabilities in their systems and ensure that data protection measures are up-to-date. This helps prevent breaches before they occur and provides assurance to players like Emma that their data is safe.
- **Access Controls:** Limiting who within the casino's organization can access sensitive player data is essential. Only authorized personnel should be able to view or manage personal information, reducing the risk of insider threats or data misuse.

Responding to Data Breaches

Despite the best efforts to secure player data, breaches can still happen. In the event of a **data breach**, casinos have an ethical and legal obligation to **notify affected players immediately**. For Emma, this means that if her data is compromised, she should receive a prompt notification explaining what happened, what data was affected, and what steps she can take to protect herself from further harm.

Casinos should also have a well-defined **breach response plan**, which includes notifying relevant authorities (such as data protection regulators) and providing support to players, such as offering identity theft protection services or credit monitoring in the event of a breach.

Player Rights

Privacy regulations such as GDPR and CCPA empower players like Emma with certain **rights over their data**. These rights are essential for ensuring that players have control over how their personal information is used and can take action if they believe their privacy is being violated.

The Right to Access

Emma has the right to know what personal data the casino has collected about her. If she requests access to her data, the casino must provide her with a copy of all relevant information, such as her gameplay history, spending patterns, and any personal details stored in their systems. This transparency ensures that Emma can verify the accuracy of the information and understand how it's being used.

The Right to Be Forgotten

Emma also has the **right to be forgotten**, which means she can request the deletion of her personal data from the casino's systems. This right is particularly important if Emma decides to stop playing at the casino or if she feels uncomfortable with the amount of data being collected about her. Casinos must comply with this request as long as the data is no longer needed for legitimate business purposes (such as fulfilling a legal obligation or preventing fraud).

The Right to Correct Inaccuracies

If Emma discovers that any of the data the casino holds about her is inaccurate, she has the **right to request corrections**. This ensures that the casino's records are up-to-date and that any personalized offers or recommendations she receives are based on accurate information.

The Right to Opt-Out

Privacy regulations also give Emma the right to **opt out of certain types of data collection or marketing practices**. For example, if Emma doesn't want the casino to use her data for targeted advertising, she should have the option to opt out of this practice without being penalized or losing access to the casino's services.

Privacy-enhancing Technologies (PETs)

To further protect player privacy, casinos can adopt **Privacy-Enhancing Technologies (PETs)**, which are tools and technologies designed to minimize the risks associated with data collection and use. These technologies can help ensure compliance with privacy regulations while giving players like Emma greater control over their personal information.

Data Anonymization and Pseudonymization

One of the most effective PETs is **data anonymization**, which involves removing or obfuscating personal identifiers from data sets, making it impossible to trace the data back to individual players. For Emma, this means that even if her gameplay data is analyzed for marketing or operational purposes, it cannot be linked directly to her identity.

Pseudonymization is another technique where personal identifiers are replaced with pseudonyms, allowing the data to be used for analysis while protecting the identity of individual players. This allows casinos to benefit from data analytics without compromising player privacy.

Differential Privacy

Another cutting-edge PET is **differential privacy**, a technique that adds statistical "noise" to data sets, making it difficult to identify individuals while still allowing for useful insights. For Emma, this means that her data can be included in aggregate analysis without the risk of her individual privacy being compromised.

Privacy Dashboards for Players

Casinos can also offer players like Emma **privacy dashboards**, where they can easily manage their data preferences, review what data is being collected, and adjust their privacy settings. This empowers players with greater control and visibility over their personal information, enhancing their sense of trust and security.

Navigating the Future of Privacy

As the gaming industry continues to evolve, **privacy** will remain a central concern for both casinos and players. New technologies, changing regulations, and growing awareness about data rights mean that casinos must stay ahead of the curve to ensure they are protecting player data and maintaining compliance with privacy laws.

The Importance of a Privacy-First Culture

For casinos to truly protect player privacy, they must adopt a **privacy-first culture**—one that prioritizes data protection at every level of the organization. This means embedding privacy considerations into the design of new games, marketing campaigns, and customer service processes. It also means training employees on the importance of data security and privacy compliance, ensuring that everyone in the organization understands their role in safeguarding player information.

Future Regulations and Compliance

As privacy regulations continue to evolve, casinos must stay informed about new developments in data protection laws. Future regulations may introduce even stricter requirements for how data is collected, used, and stored. By being proactive and ensuring that privacy is a top priority, casinos can navigate these changes while continuing to provide players like Emma with a seamless and secure gaming experience.

Key Takeaways

1. **Privacy is a Legal and Ethical Responsibility:**
   Casinos must comply with regulations like GDPR and CCPA, ensuring that player data is collected, stored, and used responsibly while respecting players' rights to privacy.
2. **Transparency Builds Trust with Players:**
   Players like Emma need to know how their data is being used and have access to clear privacy policies and settings that allow them to control their personal information.
3. **Data Security is Critical for Protecting Player Information:**
   Strong encryption, multi-factor authentication, and regular security audits are essential for safeguarding sensitive player data from breaches and cyberattacks.
4. **Player Rights Must Be Respected and Enforced:**
   Players should have the right to access, correct, delete, or opt out of the use of their personal data, empowering them with control over their information.
5. **Privacy-Enhancing Technologies (PETs) Offer Additional Protections:**
   Tools like data anonymization, pseudonymization, and privacy dashboards can help casinos ensure compliance with privacy regulations while minimizing risks to player privacy.

12.

RESPONSIBLE GAMING

Protecting Players and Ensuring Responsible Practices

Emma enjoys the thrill of the casino—whether it's playing a high-stakes game or collecting points through a loyalty program, the experience is exhilarating. But what happens when that excitement crosses a line? When gambling becomes more than entertainment and begins to impact a player's well-being, the casino has a responsibility to step in. Today, the gaming industry faces increased scrutiny over the ethics of player engagement and the role of responsible gaming in protecting players like Emma from gambling-related harm. From the adoption of self-exclusion programs to initiatives like DraftKings' $100 million responsible gaming program, the conversation around responsible gaming has never been more important.

Responsible Gaming

In an industry designed to engage and entertain, it's easy for some players to lose sight of responsible gaming practices. The casino experience is built on excitement—whether it's the flashing lights of a slot machine or the anticipation of a winning hand at the blackjack table. But for some players, like Emma, the lines between entertainment and problematic gambling can blur. It refers to the strategies, tools, and regulations that casinos implement to ensure that players engage with gaming in a way that is safe, balanced, and mindful of the risks associated with gambling. It is a shared responsibility between the player and the casino, with casinos obligated to create an environment where fun doesn't evolve into harm. With gambling addiction and problem gaming becoming serious public health concerns, responsible gaming initiatives are designed to protect players from excessive gambling, financial harm, and emotional distress.

Let's explore the critical role of responsible gaming in the casino industry, highlighting initiatives like **DraftKings' $100 million commitment to responsible gaming**, as well as the various regulations and protections put in place by **gaming commissions** worldwide. We'll also examine the growing concerns around gamification and how casinos can balance engagement and responsibility.

DraftKings and the $100 Million Commitment to Responsible Gaming

In 2022, **DraftKings** made headlines by announcing a **$100 million investment** in responsible gaming, a move that signaled a new era of accountability and awareness in the gaming industry. As one of the largest online sportsbooks and daily fantasy sports platforms in the world, DraftKings recognized the importance of ensuring that its players could enjoy gaming without falling into harmful patterns.

Key Elements of DraftKings' Initiative

The $100 million responsible gaming initiative includes several key components aimed at preventing gambling addiction, promoting player well-being, and offering support for those who may be at risk. For players like

Emma, these initiatives can provide valuable resources and tools for managing their gaming habits responsibly.

- **Educational Programs:** DraftKings has invested in educational programs designed to raise awareness about the risks associated with gambling. These programs help players like Emma understand the signs of problematic gambling and provide information on how to seek help if needed.
- **Self-Exclusion and Time Limits:** Players can set **self-exclusion** periods, during which they are barred from participating in any gambling activities on the platform. Emma can also set **time and spending limits**, ensuring that her gaming activities remain within healthy boundaries.
- **Collaborations with Research Organizations:** DraftKings partners with universities and research institutions to better understand gambling behaviors and develop more effective tools for promoting responsible gaming.
- **Tools for Managing Gaming Habits:** For Emma, DraftKings offers personalized tools that allow her to track her gaming habits, set deposit limits, and receive notifications if she's approaching her pre-set thresholds. These tools empower players to stay in control and make informed decisions about their gaming.

By making such a significant financial commitment, DraftKings not only positions itself as a leader in responsible gaming but also highlights the growing expectation that gaming operators prioritize player protection alongside profits.

The Role of Gaming Commissions in Player Protection

Gaming commissions worldwide play a crucial role in setting and enforcing **responsible gaming standards**. For Emma, these regulations provide peace of mind, ensuring that casinos are held to high standards of fairness, transparency, and player protection. Different regions have varying regulations, but the core principles remain the same: protecting vulnerable players, preventing problem gambling, and ensuring that gaming operators act responsibly.

Player Protection Rules and Compliance

Gaming commissions enforce a wide array of player protection rules that casinos must follow to maintain their licenses. These include:

- **Self-Exclusion Programs:** Many gaming commissions require casinos to offer **self-exclusion programs**, where players like Emma can voluntarily ban themselves from participating in gambling activities for a set period. These programs help prevent players from engaging in gambling during vulnerable moments and are a key component of responsible gaming.
- **Advertising Restrictions:** To prevent predatory marketing, many commissions have strict rules on how casinos can advertise to players. For instance, they may restrict targeted advertising to players who have opted out or are on self-exclusion lists. This ensures that players who have taken steps to limit their gambling are not bombarded with tempting offers.
- **Transparency in Game Design:** Gaming commissions require casinos to provide clear and

accurate information about the odds of winning, ensuring that players like Emma are fully informed before participating in any games of chance. This transparency helps prevent misconceptions about the likelihood of winning and encourages more responsible play.
- **Funding for Problem Gambling Support Services:** Many gaming commissions require casinos to contribute funds supporting research, education, and treatment programs for problem gambling. This financial support helps ensure that players who need help have access to the resources they need.

Regulatory Compliance

Failure to comply with these responsible gaming regulations can result in heavy fines, license suspensions, or even shutdowns of operations. Gaming commissions regularly audit casinos to ensure they meet the required standards. Emma's gaming experience is subject to oversight, and she can trust that the casino is operating fairly and ethically.

Gamification – Engagement vs. Responsibility

Gamification refers to the use of game-like elements—such as points, rewards, and achievements—to encourage player engagement. While gamification can enhance the casino experience for players like Emma by making gaming more interactive and rewarding, it also raises ethical concerns about how these elements might encourage excessive play.

The Risks of Gamification

The primary concern around gamification is that it can encourage **compulsive behavior**. For example, loyalty programs that reward players with points for every dollar spent or for completing specific gaming challenges might push players to continue gambling even when it's no longer fun or financially safe. For Emma, this could mean staying at the slot machines longer than planned in an effort to reach the next loyalty tier or earn a reward.

Additionally, gamification can sometimes mask the risks associated with gambling, making it feel more like a game of skill than a game of chance. By blurring the lines between gaming and gambling, casinos may unintentionally encourage irresponsible behavior.

Balancing Engagement and Responsibility

To address these concerns, casinos must strike a balance between using gamification to engage players and ensuring that they are promoting **responsible gaming** practices. Some strategies to achieve this balance include:

- **Incorporating Responsible Gaming Messaging into Gamification:** Casinos can integrate responsible gaming reminders into their loyalty programs and gamified elements. For example, Emma might receive a message reminding her to take a break or set limits as she reaches a new milestone in a loyalty program.
- **Limiting Reward Structures:** Instead of offering rewards based solely on the amount of money spent, casinos can structure their loyalty programs to reward **time spent responsibly** or **engagement with responsible gaming tools**. For instance, Emma could earn loyalty points for

setting a spending limit or taking a break after a certain amount of playtime.
- **Clear Communication about the Risks:** While gamification can enhance the player experience, it's important that casinos are transparent about the risks of gambling. This includes providing clear information about the nature of games of chance, the odds of winning, and how spending more money doesn't increase the likelihood of success.

Laws Against Exploitative Gamification

Recognizing the potential risks of gamification in gambling, several jurisdictions have introduced **laws and regulations** aimed at preventing the use of **exploitative gaming practices** that encourage addictive behavior. These laws are designed to ensure that while casinos can use gamification to enhance the player experience, they do so in a way that does not promote harmful gambling patterns.

Key Legal Protections

- **Limitations on Loyalty Programs:** Some gaming commissions have set limits on how loyalty programs can be structured to prevent them from encouraging excessive gambling. For example, casinos may be restricted from offering loyalty points based solely on the amount of money wagered and must instead promote responsible gaming behavior.
- **Ban on "Loot Box"-Style Mechanics:** In some jurisdictions, "loot box" mechanics, where players can spend money for a chance to win virtual prizes or rewards, have been regulated or banned outright. These mechanics can encourage players to spend more than intended in pursuit of valuable but random rewards, which can mirror addictive

gambling behaviors. For Emma, these protections ensure that her gaming experience is fair and not exploitative.
- **Required Responsible Gaming Tools:** Laws may require casinos to implement tools that encourage responsible gaming, such as mandatory playtime limits, deposit caps, or self-exclusion options. For Emma, this means that she always has access to tools that help her stay in control of her gaming habits.

Enforcement of Responsible Gamification Practices

Gaming commissions actively enforce these laws, ensuring that casinos design their gamification strategies in a way that protects players. Casinos that fail to comply with these regulations can face significant penalties, including fines, loss of licenses, or reputational damage.

Tools and Resources for Players

Responsible gaming initiatives offer players like Emma a range of tools and resources to help them manage their gaming habits and protect themselves from gambling-related harm. Casinos can promote a safer and more enjoyable gaming environment by making these tools readily available and easy to use.

Self-Exclusion Programs

One of the most effective tools for promoting responsible gaming is the **self-exclusion program**. For Emma, this program allows her to voluntarily exclude herself from all gambling activities for a specified period, giving her the ability to take a break if she feels her gambling is becoming problematic. Self-exclusion can be applied at a single casino or across multiple operators, depending on local regulations.

Deposit and Time Limits

Casinos can also offer players the option to set **deposit limits** and **time limits**, giving Emma control over how much money she can spend and how long she can play in a single session. These limits help prevent players from spending more than they can afford or staying at the casino longer than intended.

Responsible Gaming Education

Educational resources are another essential component of responsible gaming initiatives. Casinos can offer players like Emma access to information about the risks of gambling, signs of problem gambling, and where to find help if needed. For example, the casino might provide links to organizations like **Gamblers Anonymous** or **National Council on Problem Gambling (NCPG)**, where players can seek professional support.

The Future of Responsible Gaming

The future of responsible gaming lies in the continued collaboration between casinos, gaming commissions, and players to create an environment where gaming is both **safe and enjoyable**. As new technologies emerge and gaming experiences evolve, the industry will need to adapt its responsible gaming practices to ensure that players like Emma are protected.

Technology-Driven Solutions

Emerging technologies, such as **artificial intelligence (AI)** and **machine learning**, are opening up new possibilities for promoting responsible gaming. AI can analyze player behavior in real-time and detect early signs of problem gambling. For Emma, this means that if the system detects risky behavior—such as spending more than usual or playing for extended periods—it can intervene by offering support or recommending that she take a break.

AI-driven tools can also offer personalized, responsible gaming advice, tailoring recommendations to each player's habits and preferences. For instance, Emma might receive suggestions on how to set healthier limits based on her gaming patterns, ensuring that she remains in control.

Key Takeaways

1. **Responsible Gaming is a Shared Responsibility:**
   Casinos have an ethical obligation to protect players.
2. **DraftKings' $100 Million Commitment Sets a New Standard:**
   Industry leaders like DraftKings are investing heavily in responsible gaming, demonstrating the importance of education, self-exclusion, and personalized tools for protecting players.
3. **Gaming Commissions Enforce Player Protection Rules:**
   Regulatory bodies set strict guidelines for casinos, ensuring they offer self-exclusion programs, transparent game design, and tools that support responsible play.
4. **Gamification Must be Balanced with Responsibility:**
   While gamification enhances player engagement, it must be implemented in ways that do not encourage excessive or addictive gambling behaviors. Casinos should integrate responsible gaming tools into their loyalty programs and reward structures.
5. **New Technologies Offer Promising Solutions:**
   AI and machine learning offer powerful tools for detecting problem gambling in real time and providing personalized responsible gaming interventions for players like Emma.

13.

EFFORTLESS OPERATIONS

Optimizing Casino Operations for Success

The bright lights of the casino floor, the hum of slot machines, the clink of chips at the poker table—it all seems effortless to Emma as she enjoys her time at the casino. What she doesn't see is the complex web of operations running behind the scenes, ensuring that everything works smoothly, from game availability to customer service. For casinos, optimizing these operations is key to not only providing an exceptional player experience but also driving profitability and long-term success. In an industry that never sleeps, efficient operations mean the difference between thriving and merely surviving.

Casino operations encompass a wide range of activities, from the management of games and customer service to marketing, staffing, and even compliance with regulatory requirements. In a highly competitive industry, **operational efficiency** is essential for maximizing revenue, reducing costs, and ensuring that players like Emma have a seamless experience. Efficient operations also allow casinos to respond quickly to changes in player behavior, technological advancements, and evolving market conditions.

For Emma, efficient casino operations translate into a smooth and enjoyable experience, with minimal downtime, fast service, and a gaming environment that feels welcoming and well-maintained. For casino operators, operational efficiency means streamlining processes, optimizing resources, and using technology to automate tasks and improve decision-making.

Let's explores the key strategies and technologies that can help casinos optimize their operations, covering everything from data analytics and artificial intelligence (AI) to staff management and the integration of gaming systems. We'll also highlight how operational improvements contribute to long-term success by enhancing the player experience, boosting profitability, and reducing the complexity of managing a large-scale gaming environment.

Technology for Operational Excellence

Technology plays a critical role in optimizing casino operations. From AI-driven decision-making to automation tools, **advanced technology** allows casinos to manage their operations more efficiently and make data-driven decisions in real time. For players like Emma, this technology ensures that the games are always available, services are prompt, and the overall experience is smooth and enjoyable.

AI-Powered Operations

One of the most impactful ways technology is transforming casino operations is through the use of **artificial intelligence (AI)**. AI-powered tools can analyze vast amounts of data in real time, providing insights into player behavior, game performance, and operational efficiency. For casino operators, this means being able to make decisions faster and with greater accuracy.

For example, AI can monitor player traffic patterns and predict peak times, allowing the casino to adjust staffing levels, optimize game availability, and ensure that resources are allocated efficiently. If Emma tends to visit during specific hours, AI can predict when similar players are likely to arrive and ensure that the right number of staff members are on hand to provide high-quality service.

AI can also be used to optimize **game performance** by identifying underperforming machines and suggesting changes to improve engagement. For example, if a particular slot machine isn't attracting many players, AI can analyze the data and recommend repositioning the machine or offering targeted promotions to drive more traffic.

Automation for Streamlined Operations

Automation is another key technology that helps casinos optimize their operations. By automating routine tasks, casinos can free up staff to focus on more complex and value-added activities. For example, automated systems can handle:

Inventory management: Tracking and replenishing supplies in real-time ensures that the casino is always stocked with essential items, from chips to drinks.

Customer service inquiries: Chatbots and AI-driven virtual assistants can provide players like Emma with instant answers to common questions, reducing wait times and improving the overall experience.

Game maintenance: Automated systems can monitor the status of slot machines and table games, alerting staff when maintenance is needed and minimizing downtime.

Automation improves operational efficiency and enhances the player experience by ensuring that services are delivered quickly and consistently.

Data Analytics

Data analytics is a powerful tool for optimizing casino operations, providing valuable insights that help casinos make smarter decisions about everything from staffing to game placement. For players like Emma, data-driven operations ensure that their needs are anticipated and met, creating a more personalized and enjoyable experience.

Predictive Analytics for Staffing and Resource Allocation

One of the most important applications of data analytics in casino operations is **predictive analytics**, which uses historical data to forecast future trends. For example, by analyzing patterns in player traffic, spending habits, and game preferences, casinos can predict when certain areas of the casino will be busiest and adjust staffing levels accordingly. This ensures that there are always enough staff members available to assist players, reducing wait times and improving customer satisfaction.

For instance, if data shows that Friday nights tend to attract high numbers of slot machine players like Emma, the casino can ensure that more staff are assigned to those areas during peak times. Similarly, if certain games tend to be more popular during specific hours, the casino can adjust its floor layout to ensure that those games are easily accessible.

Predictive analytics can also help casinos manage **inventory** more efficiently. By forecasting demand for items such as drinks, chips, and promotional materials, casinos can reduce waste, minimize stockouts, and ensure that players always have access to what they need.

Optimizing Game Placement and Performance

Data analytics is also invaluable for **optimizing the placement of games** on the casino floor. By analyzing player behavior, casinos can identify which games are most popular and which areas of the casino see the most foot traffic. For Emma, this means that her favorite games are always easy to find and conveniently located.

For example, if data shows that players tend to cluster around specific types of slot machines or table games, the casino can adjust its floor layout to make those games more prominent, increasing player engagement and driving higher revenue. Conversely, underperforming games can be moved to less visible areas or replaced with more popular options.

Data analytics also helps casinos track **game performance** in real time, allowing operators to identify trends and make adjustments as needed. If a particular slot machine is performing well, the casino can promote it more heavily, while underperforming machines can be retired or replaced.

Staff Management

A well-run casino relies on its staff to provide high-quality service and maintain a smooth operation. **Staff management** is a critical component of operational efficiency, ensuring that the right people are in the right places at the right times. For players like Emma, this means receiving prompt, friendly service, whether they're ordering a drink, playing a game, or interacting with customer support.

Dynamic Staffing Models

Dynamic staffing models, powered by data analytics, allow casinos to adjust staffing levels based on real-time player demand. Instead of relying on fixed schedules, casinos can use **predictive models** to determine when more staff will be needed and when fewer employees can handle the workload. For example, if Emma visits during a particularly busy weekend, the casino's data systems can predict the influx of players and ensure that enough dealers, waitstaff, and customer service representatives are available.

Dynamic staffing not only improves the player experience by reducing wait times and ensuring consistent service but also helps casinos optimize labor costs by preventing overstaffing during slower periods.

Training and Employee Engagement

Efficient operations also depend on a well-trained, motivated workforce. Casinos that invest in **staff training** and **employee engagement** initiatives are more likely to see higher levels of productivity and better customer service. For Emma, this means that every interaction she has with staff members is positive, whether she's receiving assistance with a game or asking for information about a promotion.

Training programs should focus on equipping employees with the skills they need to provide exceptional service, handle player inquiries, and respond to issues quickly. Employee engagement initiatives, such as recognition programs or opportunities for career development, can also help keep staff motivated and committed to delivering a high-quality experience.

Integrating Systems for Seamless Operations

In a modern casino, multiple systems must work together to create a seamless experience for players like Emma. From loyalty programs to game management and payment processing, **system integration** is essential for ensuring that all aspects of the casino's operations function smoothly.

Connecting Loyalty Programs with Player Data

Loyalty programs are a key component of player engagement, and integrating them with **real-time player data** allows casinos to offer personalized rewards and promotions based on individual preferences. For Emma, this means that her loyalty points are updated instantly as she plays, and she receives targeted offers that align with her gaming habits.

System integration ensures that loyalty programs are connected to every part of the casino's operations, from the games themselves to the marketing department. This enables the casino to track player behavior in real time and offer rewards that are both timely and relevant.

Streamlining Payment Systems

Efficient payment systems are critical for ensuring that players like Emma can easily deposit funds, withdraw winnings, and manage their accounts. By integrating payment systems with other operational systems, casinos can provide a **frictionless payment experience**, whether Emma is playing at a slot machine, engaging in online gaming, or visiting a physical casino.

For example, if Emma prefers to use a digital wallet, the casino's payment system should be able to process transactions quickly and securely across multiple channels, ensuring that she can manage her funds with ease. Integrated payment systems also reduce the risk of errors, such as misapplied credits or delayed payouts, improving the overall player experience.

Compliance and Security

Casinos operate in a highly regulated environment, with strict rules governing everything from game fairness to data privacy. Ensuring **compliance** with these regulations is essential for maintaining the casino's reputation, avoiding legal penalties, and protecting players like Emma from harm.

Compliance with Gaming Regulations

To ensure fair play, casinos must comply with regulations set by **gaming commissions**, which govern everything from the randomness of slot machines to the security of financial transactions. Regular audits and inspections are conducted to verify that the casino is adhering to these rules. For Emma, this means that she can trust the casino to provide a fair and transparent gaming experience.

Operational efficiency plays a key role in maintaining compliance, as casinos must ensure that all games are functioning properly, that player data is secure, and that financial transactions are conducted in accordance with the law.

Data Security and Privacy

With the rise of digital gaming and the collection of player data, **data security** has become a critical concern for casinos. Players like Emma expect their personal information to be protected from cyberattacks and unauthorized access. Operational efficiency in this area means ensuring that robust security protocols are in place, including encryption, multi-factor authentication, and regular security audits.

By integrating data security measures into every aspect of their operations, casinos can protect both their players and their businesses from the risks associated with data breaches and privacy violations.

The Future of Casino Operations

As the gaming industry continues to evolve, the future of casino operations will be defined by **greater automation, advanced AI-driven insights, and increased integration across systems**. For players like Emma, this means an even more seamless, personalized experience, where every aspect of their interaction with the casino is optimized for convenience, engagement, and enjoyment.

AI and Machine Learning for Continuous Improvement

In the future, AI and machine learning will play an even larger role in optimizing casino operations. These technologies will enable casinos to continuously learn from player behavior and operational data, making real-time adjustments to everything from staffing levels to game availability. For Emma, this means that her experience will be constantly fine-tuned, ensuring that she always has access to the games and services she enjoys most.

Automation of Complex Processes

While automation is already streamlining many aspects of casino operations, the future will see even more complex processes being automated. From fraud detection to customer service, automation will allow casinos to operate more efficiently and at a larger scale, while maintaining a high level of service quality.

Key Takeaways

1. **Technology Drives Operational Efficiency:**
   AI, data analytics, and automation are key tools for optimizing casino operations, helping casinos make smarter decisions, allocate resources effectively, and deliver a seamless experience to players like Emma.
2. **Predictive Analytics Enhances Decision-Making:**
   By using data to predict player behavior and optimize staffing, game placement, and resource allocation, casinos can create a more engaging and profitable environment.
3. **Staff Management is Essential for Quality Service:**
   Dynamic staffing models and employee engagement initiatives ensure that players receive prompt, high-quality service, while also optimizing labor costs for the casino.
4. **System Integration is Key to Seamless Operations:**
   Connecting loyalty programs, payment systems, and gaming systems ensures that all aspects of casino operations function smoothly, providing a frictionless experience for players.
5. **Compliance and Security Must Be Prioritized:**
   Ensuring compliance with gaming regulations and protecting player data are essential for maintaining trust and avoiding legal penalties while also providing a fair and secure gaming experience for players.

14.

THE HUMAN FACTOR

Hosts Are Critical in Gaming!

Emma walks through the lively casino floor, greeted by a familiar face—a VIP host who knows her favorite games, offers personalized suggestions, and ensures her experience is top-notch. While the casino has adopted AI-powered systems to optimize games and send her tailored offers, it's the personal touch from the host that makes Emma feel truly valued. In an industry driven by technology and data, the role of the casino host remains as important as ever. But how do casinos ensure they have the best hosts? How are these hosts trained, retained, and measured in a world increasingly driven by automation and AI?

Despite the rise of AI and digital engagement, **human hosts remain essential** in the gaming industry, especially in creating strong, personal connections with high-value players like Emma. While AI can send targeted promotions, manage loyalty points, and analyze player data, hosts provide the **human connection** that builds loyalty, enhances player satisfaction, and deepens the emotional bond between the player and the casino.

Casino hosts are especially critical in maintaining relationships with **VIP and high-roller players** who expect personalized attention and tailored experiences. For Emma, the relationship she has with her host can make the difference between feeling like just another player and feeling like a valued guest. Casino hosts not only ensure that Emma's preferences are met but also anticipate her needs, creating an experience that goes beyond the game itself.

This chapter explores the critical role of casino hosts, offering strategies for **training and retaining the best hosts**, measuring their performance, and integrating AI to support and enhance the human touch they provide.

Training Casino Hosts for Success

Training is the foundation for creating **exceptional casino hosts** who can connect with players like Emma and deliver a personalized, high-quality experience. Successful hosts must possess a unique blend of skills, including excellent communication, strong interpersonal abilities, attention to detail, and a deep understanding of both the casino environment and the needs of high-value players.

Key Components of Host Training

To ensure that hosts are equipped to succeed, casinos should implement a comprehensive training program that covers the following key areas:

- **Personalization and Player Preferences:** Hosts must learn how to identify and track the unique preferences of each player. For Emma, this means understanding which games she enjoys, what times she typically visits, and what types of rewards motivate her to return. Training hosts to gather

and utilize this information helps create a more personalized experience for every player.
- **Customer Relationship Management (CRM):** Hosts need to be proficient in using the casino's CRM system to manage player data, track interactions, and follow up with players in a timely manner. Emma should feel like her host is always aware of her latest activity, whether it's a recent visit or a special request she's made. A well-trained host can use CRM tools to anticipate Emma's needs and make her feel valued.
- **Communication and Emotional Intelligence (EQ):** Building strong relationships with players requires excellent communication and emotional intelligence. Hosts should be trained to pick up on social cues, manage player emotions (especially during high-stakes games), and handle both positive and negative feedback gracefully. Emma should always feel like her host is approachable, empathetic, and ready to assist her with any concerns.
- **Understanding Player Value:** Hosts must be able to assess the value of players and prioritize their efforts accordingly. For high-value players like Emma, this might mean offering personalized rewards, exclusive access to events, or other perks that make her feel appreciated. Training hosts to identify high-value players ensures that the most important players receive the attention they deserve.

Ongoing Training and Development

Casino hosts need **ongoing training** to stay current with industry trends, changes in player behavior, and the introduction of new tools and technologies. Continuous education can include advanced CRM training, emotional intelligence workshops, and updates on responsible gaming practices. This ensures that hosts remain effective and continue to enhance the player experience over time.

Retaining the Best Hosts

Attracting and retaining top talent is essential for maintaining a successful team of casino hosts. The best hosts are those who can forge long-lasting relationships with players, driving repeat visits, loyalty, and increased spending. But how can casinos ensure they retain these valuable employees in a highly competitive industry?

Creating a Supportive and Engaging Work Environment

One of the most important factors in retaining top hosts is creating a **supportive and engaging work environment** where hosts feel valued and motivated. For Emma's host, this might mean having access to the right tools and resources to do their job effectively, as well as feeling recognized for their efforts. Key strategies include:

- **Recognition and Rewards:** Just as players like Emma appreciate rewards for their loyalty, hosts also need recognition for their hard work. Casinos should implement performance-based incentives, such as bonuses or other rewards, that encourage hosts to deliver exceptional service. Public recognition, such as "Host of the Month" awards, can also boost morale and incentivize hosts to go above and beyond.
- **Career Development Opportunities:** Providing hosts with opportunities for **career advancement** is key to retaining top talent. Hosts who feel they have a future within the casino are more likely to stay loyal to the organization. Casinos can offer training programs for advancement into

management roles or other positions within the casino, ensuring that hosts have a clear career path ahead of them.
- **Work-Life Balance:** The nature of the gaming industry can lead to long hours and high-pressure situations. Offering flexible scheduling and promoting a healthy work-life balance can help prevent burnout and ensure that hosts remain engaged and motivated.

Aligning Compensation with Performance

To retain the best hosts, casinos must offer **competitive compensation packages** that reflect the host's contribution to the casino's success. Hosts who manage high-value players like Emma, driving significant revenue and loyalty, should be compensated accordingly. Performance-based incentives can include:

- **Commission on Player Spend:** Hosts can receive a commission or bonus based on the spending or engagement levels of the players they manage. For Emma's host, this means that if she visits the casino frequently and spends more money, the host's compensation increases accordingly.
- **Loyalty and Retention Bonuses:** Casinos can offer bonuses based on the retention of high-value players, rewarding hosts for building strong, long-term relationships. For example, if Emma remains loyal to the casino over a certain period, her host might receive a bonus for successfully retaining her business.

Measuring Host Performance

To ensure that hosts are performing at their best, casinos must implement **clear metrics** to measure host effectiveness. These metrics should focus on both quantitative and qualitative aspects of the host's role, capturing not only how much revenue they drive but also how well they build relationships and enhance player satisfaction.

Key Metrics for Measuring Host Success

- **Player Engagement:** One of the most important metrics for measuring host performance is **player engagement**. This can include tracking how often players like Emma visit the casino, how long they stay, and how much they spend. Hosts who can consistently drive higher engagement from their assigned players are delivering real value to the casino.
- **Retention Rates:** Another key metric is the **retention rate** of high-value players. Hosts should be measured on how well they maintain relationships with players like Emma over time, ensuring that they return to the casino regularly. High retention rates indicate that hosts are building strong, lasting connections with their players.
- **Player Satisfaction Scores:** Measuring **player satisfaction** through surveys or direct feedback provides insight into the quality of the host's service. For Emma, this means being able to provide feedback on how well her host meets her needs, how responsive they are, and how satisfied she is with the overall experience. High satisfaction scores indicate that the host is excelling in their role.
- **Revenue Generated by Players:** While player engagement and satisfaction are critical, hosts should also be measured on the **revenue generated** by the players they manage. Hosts who are able to drive increased spending from players

like Emma are contributing directly to the casino's bottom line.
- **Number of Player Touchpoints:** Tracking the number of **touchpoints** or interactions between the host and their assigned players provides insight into how actively the host is managing relationships. This includes phone calls, emails, in-person meetings, and other forms of communication. The more engaged a host is with their players, the more likely they are to drive loyalty and retention.

Integrating AI to Support Hosts

While human hosts remain critical in the gaming industry, **artificial intelligence (AI)** can play a valuable role in supporting hosts and enhancing their ability to deliver personalized service. By integrating AI into the casino's operations, hosts can be equipped with tools that help them better understand player behavior, predict needs, and tailor their interactions with players like Emma.

AI for Personalization

AI can analyze vast amounts of player data to identify patterns and preferences, giving hosts valuable insights into how to personalize their interactions. For Emma's host, AI can track her gaming habits, spending levels, and engagement with promotions, allowing the host to offer suggestions that align with her preferences. For example, if Emma enjoys playing specific types of slot machines, AI can recommend similar games that she might find appealing, and the host can relay this information to her during their next interaction.

Predictive Analytics for Player Behavior

AI-powered **predictive analytics** can help hosts anticipate when players like Emma are likely to visit the casino or engage with certain promotions. By predicting Emma's behavior, the host can proactively reach out to her with personalized offers or invitations to special events, ensuring that she remains engaged and feels valued. For example, if AI predicts that Emma is likely to visit during a holiday weekend, the host can send her a special promotion tailored to her preferences before she even arrives.

Automating Routine Tasks

AI can also help hosts by **automating routine tasks**, freeing up their time to focus on building relationships with players. For example, AI can automatically send birthday greetings, track loyalty point balances, and provide players with updates on promotions. This allows hosts to focus on more meaningful interactions, such as personalized check-ins or invitations to exclusive events.

The Future of Casino Hosts

As technology advances, the casino host's role will evolve, but the **human touch** will remain essential. While AI can provide valuable insights and automate certain tasks, the human connection sets hosts apart and builds loyalty with players like Emma.

Combining Human and AI Strengths

The future of casino hosting lies in combining the strengths of human hosts and AI-powered tools. AI can help hosts manage large amounts of data and personalize their interactions, while hosts provide the empathy, emotional intelligence, and personal attention that AI cannot replicate.

For Emma, this means receiving the best of both worlds—a seamless, data-driven experience enriched by her host's personal touch.

Key Takeaways

1. **Training is Essential for Creating Exceptional Hosts:**
   A well-rounded training program equips casino hosts with the skills they need to personalize player experiences, build relationships, and manage player data effectively.
2. **Retaining Top Hosts Requires Recognition and Career Development:**
   Offering competitive compensation, recognition programs, and opportunities for career advancement helps casinos retain their best hosts and reduce turnover.
3. **Clear Metrics are Key to Measuring Host Performance:**
   Metrics such as player engagement, retention rates, satisfaction scores, and revenue generation provide valuable insights into how well hosts are performing in their roles.
4. **AI Enhances, But Doesn't Replace, Human Hosts:**
   AI-powered tools can support hosts by providing personalized player insights, predicting behavior, and automating routine tasks, allowing hosts to focus on building meaningful connections with players.
5. **The Future is a Balance Between Human and Technological Strengths:**
   The most successful casinos will be those that combine the personal touch of human hosts with the data-driven capabilities of AI, creating a seamless, personalized experience for players like Emma.

15.

GAMIFICATION'S EDGE

Balancing Fun and Responsibility

Emma sits at her favorite slot machine, immersed in the vibrant lights and sounds, but it's not just the game she's playing—she's earning points for every spin, working toward the next loyalty tier. She can see her progress bar rising, unlocking rewards and milestones. This is more than gambling; it's gamification in action, a dynamic way to engage players and enhance the fun of the casino experience. But behind the scenes, the casino must carefully balance this immersive engagement with a responsibility to protect players from harm. Gamification has become a powerful tool in the gaming industry, but how can casinos ensure that it enhances fun without pushing players like Emma toward risky behavior?

Gamification refers to the use of game-like elements, such as points, rewards, levels, and achievements, to enhance user engagement. It's a tool that has become increasingly prevalent in the casino industry as operators look for ways to increase player participation, retention, and satisfaction. For players like Emma, gamification transforms the gaming experience from a series of individual wagers into a broader, more interactive challenge, where every action contributes to a larger goal, whether it's advancing in a loyalty program, unlocking bonuses, or earning exclusive perks.

However, with great power comes great responsibility. While gamification can make gambling more enjoyable and exciting, it also has the potential to encourage excessive play, especially when players are motivated to chase rewards or reach the next milestone. This chapter explores the **dual-edged sword of gamification**, examining how casinos can leverage it to drive engagement while ensuring that it's implemented in a responsible and ethical way.

How Gamification Enhances Engagement

At its core, gamification taps into fundamental human motivations—achievement, competition, and reward. These motivations are particularly effective in casinos, where the thrill of winning and the desire to earn rewards can be heightened through gamified elements. For Emma, this means that every spin on the slot machine or hand at the blackjack table contributes to her overall progress, whether it's earning loyalty points, advancing to the next player tier, or completing a challenge that unlocks a bonus reward.

The Components of Gamification

Points and Rewards: One of the most common gamification elements is accumulating points or rewards for specific actions. For example, Emma might earn points for every dollar she spends, with those points contributing to free play, bonuses, or other rewards.

Levels and Progress Bars: Progress bars and levels create a sense of accomplishment by visually showing players how close they are to reaching a new milestone. For Emma, seeing her progress bar fill up as she plays encourages her to keep going in pursuit of the next reward.

Challenges and Quests: Casinos can offer **time-limited challenges** or quests, such as "play 100 hands of blackjack" or "win five times on a specific slot machine," that give players extra incentives to stay engaged and play longer. For Emma, completing these challenges unlocks exclusive rewards and makes her gaming experience feel more interactive and goal-oriented.

Leaderboards and Competitions: Leaderboards encourage competition by ranking players based on their performance or points earned. This might mean seeing Emma's name rise on a leaderboard for top players in a weekly slot tournament, pushing her to play more to maintain her position.

Gamification turns passive gambling into a **more active, engaging experience**, encouraging Emma to participate in multiple aspects of the casino's ecosystem, from playing games to engaging with promotions, checking her progress, and working toward specific goals.

The Risks of Over-Gamification

While gamification enhances the fun factor, it also poses **risks** if not carefully balanced with responsible gaming practices. The very elements that make gamification appealing—such as rewards, challenges, and progress bars—can also drive compulsive behavior, encouraging players to spend more time and money chasing rewards.

Encouraging Excessive Play

One of the main risks of gamification is that it can encourage **excessive play** by creating a sense of urgency or competition. For players like Emma, the desire to complete challenges or reach the next level can push her to play longer than she originally intended. If the rewards seem within reach, she might be tempted to keep playing even after she's spent her allotted budget.

For example, Emma might be close to reaching a new loyalty tier that offers substantial rewards, such as free spins, bonuses, or exclusive access to events. The psychological effect of being "almost there" can cause her to spend more time and money in the casino than she planned, potentially leading to problematic gambling behavior.

The Gamification Trap

Another risk of over-gamification is that players may **chase losses** in an effort to earn rewards. Gamified systems often reward players not just for winning but for participation—whether or not they win their wagers. This can create a disconnect between gambling outcomes and rewards, making players feel like they are still progressing even if they are losing money. For Emma, this might mean continuing to play after a series of losses because she's focused on earning enough points to unlock a free play bonus.

The challenge for casinos is to ensure that gamification doesn't unintentionally encourage risky behaviors or undermine responsible gaming principles.

Responsible Gamification

The key to successfully implementing gamification in casinos is finding the right balance between **engagement** and **responsibility**. Casinos must ensure that gamification enhances the player experience without pushing players toward excessive or problematic behavior. For Emma, this means enjoying the excitement of earning rewards and leveling up without feeling pressured to play beyond her means.

Setting Limits on Rewards and Challenges

One effective strategy for responsible gamification is setting **limits** on how rewards and challenges are structured. For example, casinos can:

Cap Rewards for Certain Time Periods: To prevent excessive play, casinos can limit how many points or rewards players can earn within a specific time frame. For Emma, this might mean that she can only earn a certain number of loyalty points per day, encouraging her to take breaks between gaming sessions rather than playing continuously to accumulate points.

Time-Limited Challenges: Instead of offering challenges that encourage continuous play, casinos can offer **time-limited challenges** that reward players for engaging responsibly. For example, Emma might earn extra loyalty points for visiting the casino three times over a month, rather than being rewarded for playing non-stop in a single session.

Reward Breaks and Self-Control: Casinos can incorporate features that reward players for setting **self-control limits**, such as allowing Emma to earn loyalty points for taking breaks or setting a budget for her gaming activities. This encourages responsible behavior and helps her stay in control of her playtime and spending.

Transparent Communication and Warnings

Transparency is key to ensuring that players like Emma understand how gamification works and what the potential risks are. Casinos should clearly communicate how rewards are earned, how points accumulate, and how gamified elements impact the player experience. For example, if Emma is close to reaching a new tier in the loyalty program, the casino should clearly outline how much more she needs to spend or play to unlock the reward—and emphasize the importance of gaming within her budget.

Responsible gaming warnings can also be integrated into the gamification system. If Emma has been playing for an extended period or if her spending exceeds a certain threshold, the system can send her a notification reminding her to take a break or check her spending. This helps maintain balance and encourages players to make informed decisions.

Empowering Players with Control

Giving players like Emma **control over their gamification experience** is another important way to balance fun with responsibility. Casinos can offer tools that allow players to set their own limits, track their progress, and manage their rewards in a way that aligns with their gaming habits and personal boundaries.

For example, Emma might have the option to:

- Set daily or weekly limits on how many points she can earn.
- Choose which challenges to participate in based on her preferences and comfort level.

- Track her spending and gameplay to ensure that she's staying within her budget while enjoying the gamified experience.

By empowering players with control, casinos can ensure that gamification remains a positive and engaging feature without encouraging unhealthy behaviors.

AI to Promote Responsible Gamification

Artificial intelligence (AI) can play a crucial role in balancing fun and responsibility in gamification by monitoring player behavior in real time and providing personalized interventions when necessary. For Emma, AI can help ensure that her gamified experience remains enjoyable and responsible by analyzing her gaming patterns and identifying potential risks.

AI-Driven Monitoring and Interventions

AI can analyze player data to detect signs of problematic behavior, such as sudden increases in spending, extended gaming sessions, or chasing losses. If the AI system detects that Emma is engaging in potentially risky behavior—such as playing excessively to complete a challenge or earn a reward—it can intervene by sending a personalized warning or suggesting that she take a break.

For example, if Emma has been playing for an extended period to complete a loyalty quest, the AI system might send her a message reminding her of the importance of taking breaks and playing responsibly. The system could also offer her an incentive to stop playing, such as bonus points for logging off and returning the next day.

Personalized Gamification Experiences

AI can also be used to **personalize gamification experiences** based on individual player behavior and preferences. For Emma, this means receiving challenges, rewards, and gamified elements that align with her gaming habits and risk profile. If Emma tends to play moderately, the system can offer her challenges that encourage balanced play, such as visiting the casino on specific days or engaging in multiple types of games over a set period.

By tailoring gamification to each player's behavior, AI helps ensure that the experience is engaging but doesn't push players toward risky or excessive behaviors.

Ensuring Responsible Gamification

With the rise of gamification in casinos, **regulations** have become increasingly important to ensure that these systems are implemented responsibly. Many jurisdictions have introduced rules that govern how gamification can be used in gambling environments to protect players from harm and prevent the promotion of addictive behaviors.

Compliance with Responsible Gaming Laws

Casinos must comply with **responsible gaming laws** that set clear guidelines for how gamification can be integrated into gambling experiences. These laws may include:

Restrictions on Gamified Rewards: Regulations may limit how rewards are structured to prevent players from being incentivized to spend more than they can afford. For example, loyalty programs offering spending points must ensure that players are not encouraged to overspend in pursuit of rewards.

Disclosure of Odds and Rewards: Casinos often must provide players with transparent information about the odds of winning, how rewards are earned, and the risks associated with excessive play. This helps players like Emma make informed decisions and avoid chasing unattainable rewards.

Mandatory Responsible Gaming Tools: Many jurisdictions require casinos to offer responsible gaming tools, such as self-exclusion programs, spending limits, and reality checks, that allow players to manage their gaming habits effectively.

Sample Gamification Formulas

When implementing gamification in the casino industry, it's crucial to incorporate formulas that not only boost player engagement but also promote responsible gaming. Below are sample formulas for various aspects of gamification, with provisions for responsible gaming:

1. Engagement Score (ES)

A formula to measure player engagement while participating in gamified activities.

$$ES = (P_a \times S_g \times L_b) - R_w$$

Where:

- P_a = Number of player actions taken (e.g., logins, bets, game completions)
- S_g = Score multiplier based on game level or difficulty
- L_b = Loyalty bonus coefficient for responsible gaming adherence (e.g., not exceeding playtime limits)
- R_w = Responsible gaming weight that reduces the score if problematic behavior (e.g., excessive time or spend) is detected

Application: Ensures that engagement metrics account for responsible gaming practices by rewarding players who maintain healthy gaming behavior.

2. Reward Allocation Formula (RAF)

Determines the allocation of rewards based on player activity and responsible gaming status.

$$RAF = \frac{(A_p \times G_c) + L_r}{R_g}$$

Where:

- A_p = Number of completed player activities
- G_c = Game contribution factor (e.g., high contribution for complex tasks)
- L_r = Loyalty rewards earned through consistent responsible play
- R_g = Responsible gaming compliance factor (reduces reward for violations)

Application: Rewards players who actively participate in games while following responsible gaming practices.

3. Player Wellness Index (PWI)

A measure that helps balance gamification and responsible gaming by assessing player wellness.

$$PWI = \frac{(H_t - E_t)}{P_t} \times W_c$$

Where:

- H_t = Healthy playtime threshold (e.g., max recommended hours of play per day)
- E_t = Actual playtime exceeding healthy thresholds
- P_t = Total playtime
- W_c = Wellness compliance multiplier for adherence to recommended guidelines

Application: Monitors and maintains healthy gaming habits by adjusting scores or rewards if excessive playtime is detected.

4. Responsible Gaming Compliance Score (RGCS)

A score to evaluate how well a player adheres to responsible gaming practices.

$$RGCS = \frac{S_r - V_{ng}}{S_r} \times 100$$

Where:

- S_r = Total responsible gaming actions taken (e.g., self-exclusion requests, time-outs)
- V_{ng} = Number of non-compliant gaming violations (e.g., exceeding set playtime limits)

Application: Provides a percentage score that reflects a player's responsible gaming adherence, which can be factored into loyalty programs or gamification strategies.

5. Challenge Completion Rate (CCR)

Tracks the rate at which players complete gamified challenges within responsible gaming limits.

$$CCR = \frac{C_c}{C_t} \times R_f$$

Where:

- C_c = Number of challenges completed
- C_t = Total number of challenges attempted
- R_f = Responsible gaming factor (applies penalties for challenges completed under excessive gaming conditions)

Application: Encourages players to complete challenges while maintaining responsible gaming behavior, ensuring healthy participation.

6. Gamification Loyalty Score (GLS)

Calculates a comprehensive loyalty score considering responsible gaming practices.

$$GLS = (E_s \times A_w \times G_p) - (P_o \times R_f)$$

Where:

- E_s = Engagement score from formula 1
- A_w = Average win rate in gamified activities
- G_p = Game participation coefficient
- P_o = Penalty for non-responsible gaming behaviors (e.g., repeated violations)
- R_f = Responsible gaming factor that scales penalties based on severity

Application: Rewards players who maintain consistent participation and responsible gaming habits while discouraging excessive or risky behavior.

7. Time-Controlled Reward Multiplier (TCRM)

Applies time-based bonuses that promote responsible gaming by rewarding breaks.

$$TCRM = 1 + \frac{B_t}{T_l}$$

Where:

- B_t = Break time taken between gaming sessions
- T_l = Recommended limit for continuous gaming (e.g., 2 hours)

Application: Incentivizes players to take breaks by applying a reward multiplier to their next gaming session after a responsible break period.

8. Session-Based Responsible Engagement (SRE)

Measures responsible engagement during individual gaming sessions.

$$SRE = \frac{(S_t \times A_r)}{P_s} \times R_w$$

Where:

- S_t = Total session time
- A_r = Average revenue per session
- P_s = Penalty for excessive play within the session
- R_w = Responsible gaming weight (higher for sessions within recommended time limits)

Application: Promotes balanced engagement and discourages prolonged gaming sessions by reducing the score when responsible gaming guidelines are not followed.

Incorporating Responsible Gaming Measures

Each formula above incorporates responsible gaming provisions through:

Penalties for excessive playtime: Reducing scores and rewards when players exceed healthy gaming thresholds.

Bonuses for breaks and time-outs: Encouraging breaks by applying positive multipliers or additional rewards.

Compliance multipliers: Rewarding players who adhere to responsible gaming practices with higher loyalty scores and engagement levels.

Casinos needs to foster a more engaging environment while promoting responsible gaming practices by incorporating variations of these gamification formulas. The goal is to balance enhancing player loyalty and ensuring their well-being, ultimately creating a sustainable and positive gaming experience.

Key Takeaways

1. **Gamification Enhances Engagement but Must Be Balanced with Responsibility:**
   Gamification increases player participation and enjoyment by offering rewards, challenges, and progress tracking, but it can also encourage excessive play if not carefully managed.
2. **Setting Limits on Rewards and Challenges Encourages Responsible Play:**
   Casinos can prevent players from overextending themselves in pursuit of gamified rewards by capping rewards and offering time-limited challenges.
3. **Transparency and Control Help Players Make Informed Decisions:**
   Providing players like Emma with clear information about how gamification works and giving them control over their experience helps ensure that they can enjoy the fun of gamification without falling into risky behavior.
4. **AI Supports Responsible Gamification Through Monitoring and Personalization:**
   AI can analyze player behavior in real time, identifying potential risks and providing personalized interventions to ensure that gamification remains fun and responsible.
5. **Regulations Ensure That Gamification is Implemented Ethically:**
   Compliance with responsible gaming laws and regulations is essential for preventing the misuse of gamification and ensuring that players are protected from harm.

16.

ADAPTING TO WIN

Keeping Up with Evolving Player Needs

Emma returns to the casino after several months away, and something has changed. The games are more interactive, her favorite slot machine now has personalized features, and her loyalty rewards are tailored specifically to her preferences. It's as if the casino knew what she wanted before she did. Behind the scenes, the casino has been adapting—updating its offerings, using data analytics to better understand its players, and implementing new technologies to meet evolving player expectations. In an industry that is constantly shifting, the ability to adapt to player needs is the key to winning not just Emma's loyalty, but the loyalty of every player who walks through the door.

Adaptation in the Casino Industry

In today's fast-paced gaming environment, **player preferences** are constantly evolving. New technologies, changing demographics, and shifts in entertainment trends mean that casinos must be nimble, constantly innovating to stay ahead of these changes. Players like Emma expect more than just traditional gambling—they want an immersive, personalized experience that blends entertainment, technology, and convenience.

Adapting to meet these evolving needs is critical for casinos that want to remain competitive and keep players coming back. From incorporating emerging technologies like artificial intelligence (AI) and virtual reality (VR) to offering personalized rewards and creating new types of games, casinos must be willing to **evolve** in response to player demands. This chapter explores how casinos can stay ahead by understanding their players, embracing innovation, and continuously enhancing the player experience.

Understanding Evolving Player Needs

The first step in adapting to win is **understanding what players want**. The gaming industry is no longer one-size-fits-all, and different player demographics have different expectations. For example, Emma might be looking for a seamless, tech-driven experience with personalized rewards, while other players may seek out traditional games or social interactions.

Demographic Shifts and Player Preferences

- **Millennials and Gen Z Players:** Younger generations, including **Millennials** and **Gen Z**, are increasingly becoming a dominant force in the gaming world. These players have grown up with technology, and they expect a **digital-first experience**. For Emma, this might mean integrating her casino experience with her mobile device, offering mobile gaming options, and using technology to enhance her in-person visits. These players also value **social experiences**, often preferring games that involve social elements or opportunities for competition.

- **Older Players:** On the other hand, **older players** may still prefer traditional gaming experiences, such as classic slot machines and table games. However, even this demographic is becoming more tech-savvy, and they appreciate conveniences like easy loyalty point tracking, digital payments, and personalized service. For casinos, the challenge is finding a way to cater to both ends of the spectrum without alienating any particular group.

The Shift Toward Experiences Over Gambling

As player demographics change, so do their priorities. Many players, especially younger ones, are looking for **experiences** that go beyond traditional gambling. For Emma, this might mean that the casino needs to offer more immersive entertainment, such as live shows, interactive gaming experiences, or even non-gambling activities like dining, shopping, or wellness offerings.

Casinos that can create **destination experiences**—combining gaming with lifestyle elements—are better positioned to attract a broader audience and meet the needs of players who view casinos as part of a larger entertainment ecosystem.

Embracing Technology to Stay Ahead

To keep up with evolving player needs, casinos must embrace new **technologies** that enhance the gaming experience, improve operations, and provide deeper insights into player behavior. For players like Emma, technology creates a more seamless, convenient, and personalized experience, which is essential in keeping her engaged.

Artificial Intelligence (AI) for Personalization

AI is a game-changer when it comes to understanding and adapting to player needs. By analyzing player data in real time, AI allows casinos to offer **personalized experiences** that cater to individual preferences. For Emma, this means that her favorite games are recommended to her when she logs into the casino's mobile app, and her loyalty rewards are tailored based on her recent activity and spending habits.

AI can also help predict Emma's future behavior, allowing the casino to send her personalized offers and promotions before she even thinks to ask. For example, if AI detects that Emma tends to visit the casino more often during the holiday season, it might send her a special promotion in advance to encourage her next visit.

Virtual Reality (VR) and Augmented Reality (AR) Experiences

These technologies are becoming more prevalent in the gaming world, offering immersive experiences that go beyond traditional games. For Emma, VR might allow her to enter a fully virtual casino environment, where she can interact with games and other players in a way that feels more lifelike and engaging. AR, on the other hand, could enhance her in-person experience by overlaying digital elements onto the physical casino floor, creating new interactive opportunities.

These technologies offer exciting new ways for casinos to engage with players, blending physical and digital experiences to create a more immersive, engaging environment.

Mobile Integration

It is becoming increasingly important for players like Emma, who want to be able to access their casino experience from anywhere, at any time. For Emma, this means being able to track her loyalty points, receive personalized offers, and even play her favorite games from her mobile device. Casinos that offer a seamless mobile experience are better positioned to retain tech-savvy players who expect convenience and accessibility.

Casinos can also use mobile technology to offer **location-based services**, such as sending Emma special offers when she's nearby or offering rewards for engaging with specific games or experiences on the casino floor.

Offering Personalized Rewards and Experiences

One of the most effective ways to adapt to evolving player needs is through **personalized rewards** and experiences. Players like Emma expect more than just generic promotions—they want rewards that feel tailored to their preferences and gaming habits. By leveraging data and AI, casinos can offer **hyper-personalized** experiences that deepen player engagement and build loyalty.

Loyalty Programs Tailored to Individual Preferences

Traditional loyalty programs often reward players based on how much they spend, but today's players expect more. For Emma, a loyalty program that recognizes her unique preferences—such as offering her rewards for her favorite games or inviting her to exclusive events that match her interests—creates a sense of connection and value.

Personalized loyalty programs can include:

- **Targeted offers** based on recent gameplay or spending habits. For example, if Emma has been playing a particular slot machine frequently, the casino might offer her free spins or bonuses on that machine.
- **Exclusive access** to events or experiences that align with her interests, such as VIP events, concerts, or spa services.
- **Customized tier levels** that offer rewards based on her unique play style, rather than just how much she spends.

Experiential Rewards

Beyond monetary rewards, players like Emma are increasingly interested in **experiential rewards**—unique experiences that go beyond the casino floor. This might include invitations to luxury events, backstage access to shows, or exclusive dining experiences with celebrity chefs.

By offering experiences that are tailored to individual players, casinos can create a deeper emotional connection with their guests, turning a visit to the casino into a memorable, personalized event.

Enhancing the Social Aspect of Gaming

For many players, including younger generations, gaming is no longer a solitary activity—it's a **social experience**. Casinos that can enhance the **social aspect** of gaming are better positioned to meet the needs of players like Emma, who value social interaction and community engagement.

Social Gaming and Competitions

Casinos can create more **social gaming experiences** by offering opportunities for players to compete or collaborate with others. For Emma, this might mean participating in multiplayer games, joining team-based challenges, or competing in slot machine tournaments that allow her to see how she ranks against other players.

Casinos can also enhance the social aspect by offering **leaderboards**, where Emma can track her progress and see how she compares to her peers. These social elements not only create a sense of community but also encourage friendly competition, driving engagement and loyalty.

Integrating Social Media

Another way to adapt to evolving player needs is by integrating **social media** into the gaming experience. For Emma, this might mean being able to share her gaming achievements, loyalty rewards, or event experiences with her friends on social platforms. Casinos can also use social media to engage with players outside of the casino, offering promotions, updates, and exclusive content that keeps players connected to the brand.

By fostering a sense of community and offering social experiences both in and out of the casino, operators can meet the needs of players who crave interaction and engagement.

Continuous Innovation

In the rapidly changing gaming industry, the ability to **continuously innovate** is critical for long-term success. For casinos, this means staying ahead of trends, adopting new technologies, and constantly enhancing the player experience to meet evolving expectations.

Staying Ahead of Trends

Casinos that want to remain competitive must be proactive in identifying and responding to **emerging trends** in gaming, entertainment, and technology. For example, if a new type of interactive game becomes popular, the casino should be quick to adopt it and offer it to players like Emma. Similarly, if a new form of technology—such as facial recognition or blockchain-based gaming—begins to gain traction, the casino should explore how to integrate it into their operations.

By staying ahead of trends, casinos can position themselves as **industry leaders**, attracting players who are looking for the latest and greatest gaming experiences.

Experimenting with New Offerings

Innovation requires **experimentation**. For casinos, this means being willing to try new things—whether it's introducing new types of games, offering new loyalty rewards, or creating unique experiences that differentiate them from the competition. For example, the casino might experiment with offering Emma an augmented reality treasure hunt on the casino floor, where she can earn rewards by finding hidden virtual objects.

Experimenting with new offerings not only keeps the player experience fresh and exciting but also allows casinos to learn what resonates most with their audience.

Key Takeaways

1. **Player Needs Are Constantly Evolving:**
   Demographic shifts and changing preferences mean that casinos must continuously adapt to meet the needs of different player groups, from younger, tech-savvy players to traditionalists.
2. **Technology is Key to Staying Competitive:**
   AI, VR, AR, and mobile integration are essential tools for enhancing the player experience, offering personalized rewards, and creating immersive environments that keep players like Emma engaged.
3. **Personalization Deepens Player Loyalty:**
   Offering personalized rewards, tailored experiences, and experiential incentives creates a stronger connection with players, making them feel valued and increasing their long-term loyalty.
4. **Social Experiences Are Increasingly Important:**
   Casinos that enhance the social aspects of gaming, whether through multiplayer experiences, social media integration, or competitions, are better positioned to meet the needs of players who value community and interaction.
5. **Continuous Innovation is Critical for Long-Term Success:**
   To stay ahead of the competition, casinos must be willing to innovate, experiment with new offerings, and proactively respond to emerging trends in gaming and entertainment.

17.

BEYOND THE GAMES

Adopting Gaming Strategies to the Broader Casino World

As Emma leaves the casino after a night of excitement, she finds that the experience doesn't end at the slot machines. The same thrilling strategies that made her casino visit so engaging are being applied across various outlets in the broader entertainment ecosystem—from luxury dining experiences to immersive shows and even online platforms. Gaming strategies, once confined to the casino floor, are now shaping how brands across industries engage their customers, offering personalized rewards, immersive experiences, and gamified elements that keep people coming back for more. In this chapter, we explore how the principles of casino gaming can be applied to a range of entertainment venues and beyond, creating more interactive, engaging, and profitable experiences for customers like Emma everywhere.

Gaming is no longer just about spinning reels or playing cards at the casino—it's an approach to creating **engagement** that can be applied across a variety of outlets, both within the casino environment and outside of it. The same mechanics that drive player loyalty and excitement in the casino world can be used to enhance experiences in restaurants, retail, entertainment shows, and even digital platforms. This convergence between gaming and other industries presents opportunities to **expand the gaming mindset** far beyond the traditional gaming floor.

For customers like Emma, the appeal of gaming lies in the combination of **personalization, rewards, competition, and excitement**. Applying these elements to other experiences—from dining to shopping and attending live events—can transform otherwise ordinary activities into **memorable, interactive experiences** that foster deeper engagement, loyalty, and satisfaction.

Gamifying Dining and Hospitality

Restaurants and hotels have long been part of the casino experience, but applying **gaming strategies** to dining and hospitality offers new ways to engage customers and create immersive experiences. For Emma, a visit to the casino could seamlessly transition into a gamified dining experience, where the gaming principles—such as rewards, challenges, and personalized offers—are integrated into her meal.

Personalized Dining Experiences

Much like the casino tailors rewards and promotions to Emma's gaming preferences, **personalized dining experiences** can elevate her visit to a restaurant or bar. Data collected from loyalty programs and customer profiles can be used to offer Emma tailored recommendations, special offers on her favorite dishes, or exclusive access to new menu items.

For example, suppose Emma frequents a high-end steakhouse in the casino. In that case, she might receive personalized offers based on her dining habits—such as a complimentary glass of wine with her favorite steak or early access to a new tasting menu. By integrating **personalization** into dining experiences, restaurants can make guests feel valued and create a more memorable visit.

Gamified Dining Challenges and Rewards

Applying **gamification** to dining can make the experience more interactive and fun. Restaurants within the casino could offer challenges, such as trying a specific number of new dishes over a month to unlock a special chef's table experience. Emma might also earn loyalty points for every meal she has, with additional bonuses for trying specific dishes, completing a "tasting tour" of different cuisines, or hitting milestones like her 10th visit.

This gamified approach encourages Emma to return more often, explore different menu options, and stay engaged with the dining experience even when she's not on the casino floor. By creating **rewards** and **achievements** for dining, restaurants can extend the excitement of the casino into the culinary world.

Elevating Shows and Entertainment

Casinos are already known for hosting spectacular **live shows**, concerts, and performances, but applying gaming mechanics to these events can take the experience to a new level. For Emma, attending a live show or concert could become more interactive and engaging, with gamified elements that encourage participation and offer rewards for her involvement.

Interactive Event Experiences

Imagine Emma attending a concert where she can use a mobile app to vote on the next song in the setlist, interact with other audience members, or participate in a live trivia challenge about the performers. These **interactive experiences** create a sense of community and competition, making the event more immersive and exciting.

By incorporating gaming strategies like **leaderboards** and **real-time challenges**, event organizers can engage the audience in new ways, creating an environment where the entertainment experience goes beyond passive viewing. Emma could even earn loyalty points or exclusive rewards for participating in these interactive elements, tying the experience back to her overall relationship with the casino and its offerings.

VIP and Experiential Rewards

Gaming strategies can also enhance the **VIP experience** at shows and events. For Emma, reaching a new loyalty tier in the casino's program might unlock exclusive access to backstage passes, meet-and-greets with performers, or special seating options. These **experiential rewards** create a sense of exclusivity and make Emma feel like a valued guest.

By integrating these rewards with the broader casino ecosystem, casinos can encourage Emma to engage more deeply with both gaming and non-gaming experiences, creating a holistic entertainment offering that keeps her coming back.

Retail and Shopping

Retail and shopping are natural extensions of the casino experience, and applying **gaming principles** to retail environments can drive engagement, increase customer spend, and create a more enjoyable experience for shoppers like Emma. By turning shopping into a gamified activity, casinos and retail outlets can create **loyalty-building experiences** that go beyond simply making a purchase.

Points, Levels, and Rewards for Shopping

Just as Emma earns points for playing games in the casino, she can also earn points for her purchases in retail stores within the casino complex. Every time she shops, she accumulates points that contribute to **tiered rewards**, such as discounts, free items, or exclusive shopping experiences.

Retail outlets can also create **shopping challenges** that encourage Emma to explore different stores or products. For example, a mall within the casino could offer a scavenger hunt where Emma earns points or special rewards for visiting a set number of stores, trying on a certain number of items, or making a purchase in a new category.

By gamifying the shopping experience, casinos can drive higher engagement and encourage players to **spend more time and money** in retail environments, all while making the experience more entertaining and rewarding.

Personalization and Exclusive Offers

Casinos can leverage their data on Emma's preferences to offer **personalized shopping experiences**. For example, Emma might receive tailored offers based on her past purchases, such as early access to new collections, exclusive discounts on her favorite brands, or invitations to private shopping events.

This personalized approach makes Emma feel valued as a customer and enhances her overall casino experience by creating continuity between her gaming, dining, and shopping activities. By making her shopping experience

feel personal, the casino deepens Emma's connection to the brand and increases her likelihood of returning.

Digital Engagement and Online Platforms

In today's increasingly connected world, players like Emma are not just engaging with casinos in person—they're also interacting with brands through **digital platforms**. Casinos can extend their gaming strategies beyond the physical space and into the digital realm, creating online experiences that complement in-person visits and keep players engaged even when they're not on the casino floor.

Online Loyalty Programs and Gamified Apps

Casinos can create **gamified mobile apps** that allow players like Emma to engage with the brand from anywhere. These apps can offer loyalty points, rewards, and challenges that extend beyond gaming, encouraging Emma to participate in quizzes, unlock achievements, and interact with the casino in new ways.

For example, Emma could earn points by participating in a daily trivia challenge, completing missions related to the casino's history or events, or even sharing her experiences on social media. These **digital touchpoints** keep Emma engaged with the casino even when she's not physically present, and they offer additional opportunities to earn rewards that enhance her next visit.

Virtual and Hybrid Events

Digital platforms also allow casinos to offer **virtual or hybrid events** that extend the reach of their entertainment offerings. For example, Emma might attend a virtual concert or livestreamed show from the comfort of her home, with the opportunity to participate in interactive elements or earn points for engaging with the event online.

By blending **physical and digital experiences**, casinos can create a continuous engagement loop that keeps players connected to the brand in multiple ways, enhancing loyalty and driving repeat visits.

Integrating Gaming Strategies into New Industries

The success of gaming strategies within the casino industry has inspired their application across a wide range of other industries, from **hospitality** to **fitness** and beyond. For Emma, the principles of gaming—such as earning rewards, completing challenges, and unlocking experiences—can be applied to almost any activity, creating a sense of excitement and achievement in areas she might not expect.

Gamifying Fitness and Wellness

Imagine Emma signing up for a fitness class at the casino's wellness center, where she earns points for completing workouts, hitting her fitness goals, or participating in group challenges. These points can then be redeemed for rewards, such as spa treatments, wellness products, or access to exclusive health services.

By applying gaming principles to **fitness and wellness**, casinos can turn routine activities into fun, rewarding experiences that encourage guests to stay active and engaged. This approach not only enhances the guest experience but also promotes a sense of well-being and community.

Expanding into Hospitality and Travel

The **hospitality and travel** industries are also embracing gaming strategies to enhance guest engagement. For Emma, staying at a hotel that offers a gamified loyalty program might mean earning points for each night's stay, unlocking rewards like room upgrades, free meals, or exclusive access to amenities.

Hotels can also create **experience-based challenges** that encourage Emma to explore different aspects of the property, such as completing a wellness challenge, visiting a set number of dining outlets, or participating in special events during her stay.

By gamifying the travel experience, hotels and resorts can build loyalty, drive repeat bookings, and create a sense of adventure for their guests.

Key Takeaways

1. **Gaming Strategies Extend Beyond the Casino Floor:** The principles that drive engagement in casinos—personalization, rewards, competition, and challenges—can be applied across a wide range of outlets, from dining and shopping to shows and online platforms.
2. **Personalization Enhances Every Experience:** Offering personalized rewards and tailored experiences—whether in a restaurant, a retail store, or a live event—makes customers like Emma feel valued and deepens their loyalty to the brand.
3. **Gamification Drives Engagement and Increases Spend:** Applying gamification to dining, shopping, and other activities creates a more interactive and rewarding experience, encouraging guests to spend more time and money while having fun.
4. **Digital Platforms Extend the Reach of Gaming Strategies:** Casinos can use digital platforms to engage players beyond the physical space, offering gamified apps, virtual events, and

online loyalty programs that keep players connected to the brand.

5. **Gaming Strategies Can Be Applied Across Multiple Industries:**
From fitness to hospitality, the principles of gaming—challenges, rewards, and competition—can be used to enhance customer experiences in nearly any industry, driving engagement, loyalty, and satisfaction.

18.

LEARNING FROM OTHERS

Cross-Industry Insights for the Future of Casinos

Emma has always enjoyed her casino visits, but lately, she's noticed something different. The games are more interactive, the loyalty rewards feel more personal, and even the dining and shopping experiences are more engaging. What she doesn't realize is that the casino is borrowing strategies from industries outside of gaming—learning from retail, tech, hospitality, and more to enhance the overall experience. As casinos look toward the future, they are increasingly turning to insights from other industries to stay competitive, innovate, and deliver the kind of personalized, immersive experience that keeps players like Emma coming back.

In the rapidly evolving world of gaming and entertainment, **cross-industry learning** is more important than ever. Casinos are no longer isolated in their approach to customer engagement—they are actively looking to **other industries** for inspiration and adopting successful strategies from retail, hospitality, tech, and beyond. This ability to learn from others has become a critical driver of innovation and growth in the casino industry.

By examining how different industries approach challenges such as **customer loyalty, personalized service, operational efficiency, and technological innovation**, casinos can find new ways to **elevate the player experience**. Whether it's incorporating retail's mastery of customer personalization, hospitality's focus on guest satisfaction, or tech's relentless innovation, casinos can apply these lessons to create a more engaging, efficient, and future-focused environment.

This chapter explores the **cross-industry insights** that casinos can adopt to shape the future of gaming, highlighting how strategies from retail, technology, hospitality, and other sectors can help casinos thrive in an increasingly competitive and tech-driven world.

Personalization from Retail – Know Your Customers

Retailers have long been experts in understanding and catering to their customers' preferences, using data and technology to deliver personalized experiences that drive loyalty and repeat business. For casinos, learning from retail's success in **personalization** can be a game-changer, especially when it comes to player engagement and satisfaction.

Data-Driven Personalization

In the retail industry, customer data is used to create **tailored experiences** that match individual preferences. Retailers track everything from shopping habits to purchase history, allowing them to send targeted offers and recommendations to each customer. For Emma, this might mean receiving an email from her favorite clothing store with a personalized discount on items she frequently buys.

Casinos can adopt a similar approach, using player data to **personalize every aspect of the gaming experience**. For Emma, this means receiving personalized offers based on her favorite games, preferred visit times, and spending habits. Whether it's tailored loyalty rewards, personalized promotions, or special invitations to events that align with her interests, this level of personalization makes her feel valued and deepens her connection to the casino.

By borrowing retail's data-driven personalization strategies, casinos can ensure that every interaction feels custom-made for players like Emma, driving higher engagement, loyalty, and satisfaction.

Seamless Omnichannel Experiences

Retailers have also mastered the art of **omnichannel experiences**, creating seamless transitions between online and offline interactions. For Emma, this might mean browsing products online, checking availability in-store, and completing her purchase either digitally or in person—all while earning loyalty points that sync across every channel.

Casinos can apply this omnichannel approach by integrating **online gaming platforms**, mobile apps, and in-person casino experiences into a unified system. For example, Emma can check her loyalty points, receive personalized offers, and even play casino games on her mobile device, with everything syncing to her in-person account when she visits the casino. This creates a more connected experience, allowing Emma to engage with the casino whenever and wherever she chooses.

Hospitality's Focus on Customer Experience

The **hospitality industry** excels at creating memorable guest experiences, with a focus on personalized service, attention to detail, and customer satisfaction. Casinos can learn from hospitality's approach to delivering exceptional service, ensuring that every player feels like a VIP, regardless of their spending habits.

VIP Service for Every Player

In hospitality, every guest is treated with care, whether they're staying in a luxury suite or a standard room. For casinos, this means offering **VIP-style service** to every player, from high-rollers to casual visitors like Emma. This doesn't necessarily require lavish perks or expensive offerings—it's about creating a **personal connection** and making players feel valued.

For Emma, this could mean having a casino host greet her personally when she arrives, remembering her favorite games, and offering her a complimentary drink based on her preferences. It's the small touches that create a sense of belonging and make her feel like she's getting special treatment, even if she's not a high-stakes player.

By borrowing hospitality's focus on personalized service, casinos can create an environment where every player feels like a VIP, increasing player satisfaction and loyalty.

Loyalty Programs and Guest Retention

Hospitality brands have perfected the art of **loyalty programs** that go beyond just rewarding frequent stays. Many hotel chains offer tiered loyalty programs that provide exclusive benefits at every level, from free upgrades to personalized services. For example, Emma might earn points during her hotel stays, eventually unlocking perks like complimentary breakfasts, spa access, or room upgrades.

Casinos can implement similar **tiered loyalty programs** that offer meaningful rewards at every level, keeping players engaged and motivated to return. For Emma, this might mean earning points for every dollar she spends on games, dining, or shopping within the casino complex, eventually unlocking special experiences like access to exclusive events, free play bonuses, or personalized gifts.

By aligning their loyalty programs with hospitality's focus on guest retention and satisfaction, casinos can create deeper connections with players and encourage long-term loyalty.

The Power of Constant Evolution

The **technology industry** thrives on **innovation**, constantly pushing the boundaries of what's possible through new products, services, and experiences. Casinos can look to the tech world for inspiration, adopting a mindset of continuous improvement and innovation to stay ahead of competitors and meet evolving player needs.

Embracing Artificial Intelligence (AI)

One of the most transformative technologies emerging from the tech industry is **artificial intelligence (AI)**, which can analyze vast amounts of data in real time, predict trends, and personalize experiences on a large scale. Tech companies like Amazon and Netflix use AI to recommend products and content to users based on their browsing and purchasing habits.

For casinos, AI offers the opportunity to **personalize the player experience** in new ways. By analyzing Emma's gaming patterns, spending behavior, and preferences, AI can predict what types of games she's likely to enjoy, when she's most likely to visit, and what promotions will resonate with her. This allows the casino to deliver highly targeted offers that feel relevant and timely.

AI can also help optimize **casino operations**, such as managing staffing levels based on real-time player traffic, identifying potential bottlenecks in service, and even predicting which games are underperforming and need to be replaced or updated.

Continuous Digital Transformation

The tech industry is built on **digital transformation**, constantly upgrading and evolving products to improve user experiences. Casinos can learn from this relentless pursuit of innovation by ensuring their digital platforms—from mobile apps to online gaming sites—are always evolving to meet player expectations.

For Emma, this might mean that the casino's mobile app regularly offers new features, such as interactive challenges, live gaming events, or integrated social media functionality. By continually improving the **digital experience**, casinos can keep players engaged even when not on the gaming floor, ensuring they remain connected to the brand.

Focus on Immersive Experiences

The **entertainment industry** has always been about creating immersive, engaging experiences that capture the attention of audiences. Whether through blockbuster movies, live concerts, or virtual reality (VR), entertainment brands know how to **create memorable moments** that keep people coming back for more.

Creating Immersive Environments

In recent years, there has been a growing focus on creating **immersive environments** that blur the lines between reality and fantasy. For example, theme parks like Disney and Universal have developed highly detailed worlds that allow guests to step into the stories they love, interacting with characters and settings as if they were real.

Casinos can apply this concept by creating **immersive gaming environments** that go beyond the traditional gaming floor. For Emma, this might mean entering a section of the casino designed to look and feel like a specific theme or era, such as a 1920s speakeasy or a futuristic space station. Interactive elements, such as augmented reality (AR) experiences or live actors playing roles in the environment, could enhance the immersion, making Emma feel like she's part of a larger story.

By creating these immersive, themed environments, casinos can transform the gaming experience into something more akin to an **entertainment spectacle**, appealing to players who are looking for more than just a chance to win money.

Expanding Beyond Gambling with Entertainment-Driven Experiences

The entertainment industry also offers lessons in **diversification**. Just as entertainment brands offer more than movies or concerts (think merchandise, streaming services, theme parks), casinos can expand their offerings to include more **entertainment-driven experiences** that appeal to a broader audience.

For Emma, this might mean being able to attend live shows, interactive performances, or virtual reality experiences within the casino complex. By incorporating a wide range of **non-gaming entertainment options**, casinos can attract new players, retain existing ones, and create a destination experience that goes beyond gambling.

Cross-Industry Partnerships

As casinos learn from other industries, they are also exploring **cross-industry partnerships** to offer players more comprehensive, multi-faceted experiences. By collaborating with brands in retail, technology, entertainment, and hospitality, casinos can create **ecosystems** that enhance the overall player journey.

Collaborations with Retail and Tech Brands

For example, casinos could partner with **retail brands** to offer in-casino pop-up shops or exclusive merchandise that ties into the casino's loyalty program. Emma might earn points for making a purchase at a luxury store within the casino, or she could redeem her loyalty points for high-end goods that aren't available elsewhere.

Similarly, partnering with **tech brands** allows casinos to integrate cutting-edge technology into the player experience. For example, the casino could collaborate with a tech company to offer exclusive VR experiences or bring in new gaming innovations that appeal to tech-savvy players like Emma.

Creating Holistic Entertainment Destinations

Casinos can also partner with **entertainment companies** to offer live performances, exclusive events, or even co-branded experiences. For example, Emma might attend a concert at the casino that's part of a larger entertainment brand's tour, or she could participate in an interactive event that blends gaming with live entertainment.

These cross-industry partnerships allow casinos to expand their offerings and create a **holistic entertainment destination** that appeals to many players, from gamers to shoppers to concertgoers.

Key Takeaways

1. **Personalization is Key to Player Engagement:**
   By learning from the retail industry's data-driven personalization strategies, casinos can offer tailored experiences that make players feel valued and deepen loyalty.
2. **Customer Experience is Paramount:**
   Borrowing from the hospitality industry, casinos should focus on delivering VIP-level service to every player, creating memorable experiences that go beyond the gaming floor.
3. **Innovation is Essential for Staying Competitive:**
   The technology industry's focus on continuous innovation can inspire casinos to adopt AI, optimize operations, and enhance the digital player experience, keeping them ahead of competitors.
4. **Immersive Experiences Drive Engagement:**
   Entertainment brands offer valuable lessons in creating immersive environments and expanding beyond traditional offerings, allowing casinos to create destination experiences that attract a wider audience.
5. **Cross-Industry Partnerships Offer New Opportunities:**
   Collaborating with retail, tech, and entertainment brands allows casinos to create comprehensive ecosystems that offer more than just gaming, enhancing the overall player journey and driving long-term loyalty.

19.

MARKETING REINVENTED

Innovations in Casino Marketing

As Emma scrolls through her phone, a personalized notification from her favorite casino pops up: a special offer just for her—free spins on a new slot machine she's been eyeing, plus an invitation to an exclusive event happening next week. The offer feels tailored to her interests and habits, seamlessly delivered through an app she uses every day. This isn't traditional casino marketing; this is the future, where data-driven insights, real-time engagement, and innovative campaigns reshape how casinos connect with players like Emma. In a world of digital transformation, casinos must reinvent their marketing strategies to remain competitive, personalized, and immersive.

The casino industry is undergoing a **marketing transformation** driven by evolving player expectations, advancements in technology, and the increasing importance of personalization. Traditional methods of mass marketing—like billboards, direct mail, and email blasts—are no longer enough to capture the attention of modern players. For customers like Emma, who are accustomed to real-time, personalized digital experiences, the casino's marketing efforts must feel relevant, engaging, and tailored specifically to their preferences.

Today's most successful casinos are shifting away from one-size-fits-all campaigns and embracing **innovative, data-driven marketing strategies** that focus on understanding each player's behavior, predicting their needs, and delivering customized offers at just the right moment. This chapter explores the innovative marketing strategies that are redefining how casinos attract, engage, and retain players, from personalized promotions to real-time engagement and cross-platform integration.

Data-Driven Personalization

Personalization has become the cornerstone of modern casino marketing, with **data-driven insights** allowing casinos to create targeted, relevant offers for individual players. Gone are the days when casinos would send the same promotion to thousands of players, hoping a few would respond. Now, marketing is **laser-focused** on delivering the right message to the right player at the right time.

Analyzing Player Data for Personalization

For players like Emma, who expect brands to know their preferences, **data analytics** plays a critical role in driving engagement. By analyzing player behavior—such as which games Emma plays most often, how much she spends, when she prefers to visit, and what promotions she's responded to in the past—the casino can develop highly personalized marketing campaigns that resonate with her specific interests.

For example, if the casino's data shows that Emma enjoys playing slot machines late in the evening, they might send her a personalized offer for free spins during that time, knowing she's likely to respond. Similarly, if Emma has a history of participating in special events, she might receive an invitation to an exclusive VIP night or a private concert.

Behavioral targeting allows casinos to go beyond generic offers and deliver highly relevant content that makes players feel valued, increasing the likelihood of engagement and loyalty.

AI-Powered Recommendations

Artificial intelligence (AI) has taken personalization to the next level by enabling casinos to predict player behavior and offer real-time recommendations. For Emma, this means receiving offers for games she hasn't tried yet but that the AI system predicts she'll enjoy based on her past behavior and the preferences of similar players.

AI-powered systems can also track Emma's interactions with the casino in real-time, delivering promotions, rewards, or personalized messages while she's on the gaming floor or using the casino's mobile app. Suppose Emma is playing a particular game and the system detects that she's about to hit a milestone. In that case, it might send her a notification offering bonus points or free spins to encourage her to keep playing.

By leveraging AI, casinos can deliver **hyper-personalized experiences** that feel timely and relevant, creating a deeper connection between the player and the brand.

Real-Time Marketing

It is one of the most exciting innovations in casino marketing, allowing casinos to engage with players while they are actively interacting with the brand. For Emma, this means receiving offers or messages based on her immediate actions, whether she's playing a game, attending an event, or browsing the casino's mobile app.

Location-Based Offers

One of the most effective real-time marketing strategies is **location-based marketing**, which uses geolocation data to send personalized offers or notifications when a player is physically near the casino or specific areas within the property. For example, if Emma is near the casino's entrance or dining area, she might receive a special offer for a free drink or a discount on a meal.

By sending targeted promotions based on Emma's location, the casino can create timely, relevant engagement that enhances her experience and encourages her to participate in additional activities.

Event-Based Engagement

Real-time marketing also extends to **event-based engagement**, where the casino sends offers or messages based on what's happening in the moment. For instance, if Emma is playing a specific game and hits a winning streak, the casino might send her a bonus offer to keep the momentum going. Or, if she's participating in a slot tournament, she might receive real-time updates on her progress and special rewards based on her performance.

This type of **dynamic marketing** keeps players like Emma engaged during key moments, enhancing the sense of excitement and making the overall casino experience more immersive.

Cross-Platform Integration

Today's players engage with casinos across multiple platforms, from in-person visits to mobile apps, websites, and social media. To maximize engagement, casinos must adopt a **cross-platform marketing strategy** that creates a seamless, integrated experience for players like Emma, whether she's on the gaming floor or at home browsing her phone.

Omnichannel Loyalty Programs

For Emma, the casino's **loyalty program** is likely the most important part of her relationship with the brand, and it must be accessible across all channels. By integrating the loyalty program with mobile apps, websites, and in-person experiences, casinos can ensure that Emma can easily track her points, redeem rewards, and receive personalized offers no matter where she is.

For example, Emma might earn loyalty points during an in-person visit, which she can then redeem for online rewards through the casino's app. Similarly, if Emma receives a promotional offer on her phone, she should be able to use it seamlessly during her next visit to the casino, without needing to print or carry anything.

Creating a **unified loyalty experience** across all platforms helps ensure that players remain engaged with the casino, whether they're playing online, attending events, or visiting in person.

Social Media and Influencer Marketing

Social media is another key platform for engaging modern players, offering casinos the opportunity to build relationships with players like Emma through **interactive**

content and influencer partnerships. For example, Emma might follow her favorite casino on Instagram or TikTok, where she can see exclusive content, participate in contests, or interact with the brand's social media influencers.

Influencer marketing is becoming increasingly important for casinos, as players look to trusted voices for recommendations and experiences. By partnering with influencers who align with the casino's brand, casinos can reach new audiences, promote special events, and create **buzz** around the brand.

For Emma, seeing her favorite influencer attend an exclusive casino event or share behind-the-scenes content can make her feel more connected to the brand and increase her likelihood of visiting the casino herself.

Gamification in Marketing

Gamification has become a powerful marketing tool, turning customer engagement into a game-like experience where players are rewarded for participating in challenges, completing tasks, or reaching milestones. Gamification makes the marketing experience more interactive, enjoyable, and rewarding for players like Emma.

Challenges and Rewards

Casinos can create **gamified marketing campaigns** that encourage players to engage with the brand through challenges, tasks, or competitions. For example, Emma might participate in a week-long challenge where she earns points for playing specific games, attending events, or visiting different areas of the casino. At the end of the challenge, Emma could redeem her points for exclusive rewards, such as free play bonuses, VIP access, or personalized gifts.

By incorporating **challenges** and **achievements** into their marketing strategy, casinos can create a more engaging experience that keeps players like Emma excited and motivated to participate.

Loyalty Tiers and Badges

Gamified loyalty programs also offer **tiered rewards** and **badges** that players can unlock as they progress through different levels. For Emma, this might mean earning a badge for visiting the casino a certain number of times or unlocking a new loyalty tier that grants her access to exclusive events or higher-value rewards.

By making the loyalty program more dynamic and game-like, casinos can encourage players to **strive for higher levels**, increasing their engagement and long-term loyalty.

Content Marketing and Storytelling

One of the most innovative trends in casino marketing is the use of **content marketing and storytelling** to build a narrative around the brand, creating an emotional connection with players like Emma. Rather than simply promoting offers or events, casinos are now telling stories that resonate with their audience, from behind-the-scenes looks at how the casino operates to player success stories and the history of the brand.

Creating Immersive Content

For Emma, engaging with a casino's content goes beyond just looking at ads—it's about experiencing the brand in new ways. Casinos can create **immersive content** that draws players into the brand's world, such as video series, podcasts, blog posts, or social media campaigns that explore different aspects of the casino experience.

For example, the casino might produce a series of short videos that showcase the excitement of a slot tournament, interview players about their favorite games, or offer behind-the-scenes looks at exclusive events. This type of content not only entertains but also **builds a connection** between the player and the brand, making Emma feel more invested in the casino's story.

Highlighting Player Success Stories

Storytelling can also focus on players themselves, highlighting their experiences and success stories. For Emma, seeing a fellow player win big or participate in a special event might inspire her to get more involved with the casino. By sharing player stories through social media, blogs, or video content, casinos can create a sense of community and inspire others to participate.

Key Takeaways

1. **Personalization is the Future of Casino Marketing:**
   Data-driven insights and AI-powered recommendations allow casinos to deliver highly personalized offers and experiences, making players like Emma feel valued and increasing engagement.
2. **Real-Time Marketing Enhances Engagement:**
   Location-based offers and event-driven messaging keep players engaged in the moment, offering timely rewards and promotions that enhance the casino experience.
3. **Cross-Platform Integration Creates a Seamless Experience:**
   By integrating loyalty programs and marketing efforts across in-person, online, and mobile platforms, casinos can create a unified experience that keeps players engaged no matter where they are.
4. **Gamification Turns Marketing into an Interactive Experience:**
   Gamified challenges, rewards, and loyalty tiers make the marketing experience more enjoyable and motivate players like Emma to participate in competitions and strive for higher rewards.

5. **Content Marketing Builds Emotional Connections:** Storytelling and immersive content allow casinos to connect with players on a deeper level, creating a narrative around the brand that resonates with players and builds long-term loyalty.

20.

THE NEXT FRONTIER

The Future of Gaming Technology

Emma steps into a casino in 2030, but this is not the typical casino she's been to before. The environment around her pulses with energy, colors shift dynamically based on her preferences, and holograms greet her as she walks through the entrance. A game table materializes in front of her, blending the physical and digital in a seamless, immersive experience. Players around her are not just interacting with cards and chips—they're exploring mixed-reality worlds that transport them into adventures beyond imagination. This isn't just gaming; it's a glimpse into the future, a new frontier where the lines between real and virtual blur, and where casinos must reimagine themselves to remain relevant in a world defined by **Vision 2030** *and beyond.*

As we look toward **Vision 2030** and beyond, the future of gaming technology promises to transform the casino experience in ways that may seem like science fiction today. For players like Emma, the casino will become more than just a place to play games—it will be an interactive, immersive environment where real and virtual worlds blend into a seamless experience, offering **holograms**, **mixed reality**, and unprecedented levels of personalization.

The casino industry is at a turning point, driven by the **demographic shift** toward younger, tech-savvy generations who expect more than just traditional gaming. These players, raised on video games, social media, and digital experiences, are looking for entertainment that is **immersive, social, and interconnected**. At the same time, cities like **Las Vegas** are evolving into global entertainment capitals, becoming the home of not just casinos but also **sports**, concerts, and major events. As the world's attention shifts to Las Vegas as a hub of entertainment, casinos must embrace this future and position themselves as leaders in a rapidly changing industry.

The Blurring of Reality

The casino of the future will transcend the boundaries between the **physical and digital worlds**, offering **holographic and mixed-reality** experiences that bring gaming to life in ways that were once unimaginable. In this world, the lines between reality and virtual space will become as fluid as the iconic scenes in *The Matrix*, where characters seamlessly transition between the real and virtual worlds. This merging of realms will reshape how players like Emma experience the casino environment.

Holograms and Interactive Environments

Imagine walking into a casino where **holograms** greet you at the door, serving as virtual hosts or even celebrity figures who guide you through the gaming experience. For Emma, her favorite artist or athlete might appear as a holographic avatar, leading her to a personalized gaming table or offering her real-time updates on upcoming events. These holograms could even play roles in immersive, narrative-driven games that turn the casino floor into a **living, interactive environment**.

Interactive environments will adapt dynamically to player preferences. Emma could choose to play a blackjack game on a holographic table that transforms into a futuristic space station or a medieval castle, depending on her mood. These environments will blend physical touchpoints (like cards or dice) with virtual elements, allowing players to experience gaming worlds that go far beyond the constraints of traditional tables and slots.

Mixed Reality – The Game Beyond the Screen

Mixed reality (MR) will take this concept even further, allowing Emma to interact with both physical and virtual elements simultaneously. MR headsets or augmented reality (AR) glasses could let her see a game playing out in the real world while adding digital layers of information or characters. For example, Emma might be playing a slot machine where the reels are real, but the entire room transforms around her based on her progress in the game. Characters from the game might appear in her field of vision, reacting to her wins or losses, adding a **cinematic storytelling element** to her gaming experience.

For casinos to **lead the charge in Vision 2030**, they will need to embrace **MR gaming** to offer a multisensory experience that combines the tactile satisfaction of real-world play with the limitless possibilities of the digital realm.

The Evolution of Player Engagement

In the future, the casino experience will be tailored to each player's unique preferences and behaviors in ways that are far more advanced than today's loyalty programs. Powered by **artificial intelligence (AI)** and **big data**, hyper-personalization will drive every aspect of the player journey—from the moment Emma steps into the casino to the offers she receives after her visit.

AI-Driven Customization

Imagine a world where AI systems can predict Emma's gaming preferences, favorite environments, and optimal playing times with near-perfect accuracy. The casino will know not only which games Emma enjoys but also when she prefers to play, how long she usually stays, and what types of rewards motivate her. AI will generate **customized offers** that are tailored to her mood in real time—if Emma is feeling adventurous, she might be invited to try a new interactive mixed-reality game, while if she's looking for relaxation, the casino could offer her a personalized spa package.

This level of **AI-driven customization** will be so sophisticated that Emma won't have to seek out new experiences—the casino will anticipate her needs and adapt to her in real time. Personalized holographic hosts, tailored events, and rewards based on behavioral analysis will become the norm, creating a highly individualized experience for each player.

Immersive Loyalty Programs

Loyalty programs will also evolve into **immersive experiences**, offering players far more than just points and perks. Emma's loyalty status might be represented by a **digital avatar** that she can customize based on her progress within the casino ecosystem. As she reaches new tiers, her avatar could unlock special abilities or exclusive access to certain areas of the casino, such as private gaming lounges or VIP events. These digital avatars could even interact with other players in the virtual space, creating a

social gaming community where loyalty is expressed through personalized, interactive elements.

The integration of **blockchain technology** could allow Emma to carry her loyalty status across multiple platforms, not just within a single casino but throughout an entire network of interconnected gaming and entertainment venues, giving her access to exclusive experiences around the world.

The Rise of Social and Competitive Gaming

One of the biggest demographic changes shaping the future of gaming is the rise of **social and competitive gaming**. Younger generations, particularly **Millennials** and **Gen Z**, are drawn to experiences that are social, interactive, and competitive, influenced by their engagement with video games, eSports, and social media platforms. For casinos to thrive beyond 2030, they must embrace this **competitive spirit** and incorporate it into their gaming and entertainment offerings.

Social Media Integration and Live Streaming

By 2030, the integration of **social media** and **live streaming** will be essential to the gaming experience. Players like Emma will expect to share their experiences in real time with friends, followers, and communities around the world. Imagine Emma live-streaming her gaming session directly from the casino floor, where her friends can interact with her, cheer her on, and even place bets in real-time on her performance. These social interactions could be integrated into the gaming platform itself, with leaderboards, achievements, and rewards that encourage competition and community engagement.

Influencer marketing will also play a key role, with influencers creating content around their casino experiences, hosting live events, and promoting unique gaming opportunities. By partnering with social media influencers, casinos can expand their reach and attract a new generation of players who are more interested in the **social aspects of gaming** than traditional solo play.

eSports and Competitive Tournaments

As **Las Vegas evolves into the sports capital of the world**, competitive gaming, particularly **eSports**, will become an integral part of the casino ecosystem. Emma could participate in **live gaming tournaments** that are streamed globally, with massive audiences watching her compete for prizes and status. Casinos will need to build **eSports arenas** and create tournament-style events that appeal to competitive gamers, blending the excitement of traditional sports with the high-stakes world of gaming.

By investing in **competitive gaming infrastructure**—from dedicated tournament spaces to live broadcasting capabilities—casinos can position themselves as leaders in the emerging world of eSports and social gaming.

Seamless Integration of Sports and Entertainment

The future of casinos will not just be about gaming; it will be about creating **multidimensional entertainment hubs** that combine sports, music, dining, and immersive experiences into a seamless offering. As Las Vegas becomes the world's premier destination for sports, casinos will need to integrate **sports betting, live events, and real-time gaming** into a single, interconnected experience.

Sports Betting and Real-Time Interactions

Sports betting is already a significant driver of revenue, but by 2030, it will become even more integrated with the casino experience. Imagine Emma placing a bet on her favorite football team while sitting at a table game, with real-time updates and odds appearing in her mixed-reality display. As the game unfolds, she can interact with live stats, make in-game bets, and even compete against other players in real-time prediction challenges.

Casinos will also host **live sports events** that go beyond traditional betting. These venues will offer immersive viewing experiences, with holographic replays, player stats displayed in real time, and the ability to switch perspectives as if you were on the field. This blending of live sports, entertainment, and gaming will create a more dynamic, interactive environment that attracts fans of all kinds.

Music, Entertainment, and Beyond

Casinos will also become centers of **live entertainment** and culture, offering concert venues, interactive shows, and music festivals that seamlessly integrate with gaming. For Emma, attending a concert at the casino won't just be about the music—it will be an interactive experience where she can participate in live challenges, earn rewards, and engage with the artists through virtual meet-and-greets or holographic performances.

This **integration of music, sports, and gaming** will make the casino a destination for entertainment in all its forms, creating a one-stop destination for immersive, multi-sensory experiences.

Preparing for the Next Frontier

For casinos to thrive in **Vision 2030 and beyond**, they must embrace a **future-forward strategy** that leverages cutting-edge technology, immersive experiences, and cross-industry collaboration. This is not just about adopting new tools—it's about **redefining the entire casino experience** to appeal to a new generation of players who expect more than just games. They want stories, social interaction, competition, and the blending of the real and virtual in ways that challenge their imagination.

Becoming an Outlier

To become an **outlier** in the casino industry, operators must adopt a **blueprint for innovation** that includes:

1. **Investing in Immersive Technology:** Embrace holograms, mixed reality, and AI to create personalized, dynamic environments that adapt to player behavior in real time.
2. **Expanding Social and Competitive Gaming:** Build infrastructure for eSports, social media integration, and live streaming to attract younger players who crave competition and community engagement.
3. **Blending Real and Virtual Worlds:** Create seamless transitions between physical and digital experiences, using storytelling and gamification to make every moment feel immersive.
4. **Integrating Entertainment and Sports:** Become a hub for sports, music, and live entertainment, offering a holistic experience that combines gaming with real-time interactions, events, and shows.

5. **Creating a Vision Beyond 2030:**
   Think beyond traditional gaming. Casinos must position themselves as **global entertainment destinations**, where every visit offers something new, unexpected, and unforgettable.

Key Takeaways

1. **The Future of Gaming is Immersive:**
   Casinos must embrace holograms, mixed reality, and AI to create dynamic, personalized experiences that blur the lines between the real and virtual worlds.
2. **Hyper-Personalization Will Redefine Player Engagement:**
   AI-driven marketing, loyalty programs, and real-time customization will create an environment where every player's experience is uniquely tailored to their preferences.
3. **Competitive and Social Gaming are the New Frontier:**
   By integrating eSports, live streaming, and social media, casinos can attract younger, competitive players who thrive in interactive, community-driven environments.
4. **Sports and Entertainment Will Become Central to the Casino Ecosystem:**
   As Las Vegas becomes the global sports capital, casinos must seamlessly integrate sports betting, live events, and entertainment to offer a multi-dimensional experience.
5. **To Lead the Future, Casinos Must Think Beyond Games:**
   Casinos of the future must become immersive entertainment hubs, blending gaming, sports, music, and digital experiences into a unified offering that appeals to the next generation of players.

21.

THE CALL

Building Tomorrow's Casino Today

Imagine stepping into a casino in 2030. It's not just a place to gamble anymore—it's an experience that surrounds you, tailors itself to your desires, and transports you to real and virtual realms. Holograms greet you at the door, the environment shifts dynamically with your mood, and your loyalty program isn't just points—it's a fully interactive experience. This isn't some far-off dream—it's the casino of tomorrow, and it's being built today by those bold enough to answer the call. The question is: Will you be one of them?

As we conclude this journey into the future of casinos, we arrive at a critical moment. The **casino industry of tomorrow** will not be built by those who merely watch the future unfold—the outliers will shape it. **These innovators** step forward and create something truly extraordinary. The future is calling, and it demands bold leadership, visionary thinking, and a commitment to **building tomorrow's casino today**. The question that every casino leader must ask themselves is: *Are we ready?*

The Need for Bold Leadership – Becoming an Outlier

In a world where every industry is racing toward digital transformation and technological innovation, standing still is no longer an option. To **win in the future**, casino operators must become **outliers**—those who are willing to break from the crowd and set a new path. But what does it mean to be an outlier in the gaming industry? How can your casino stand out in a landscape that's rapidly changing, where every player expects **more personalization**, **more immersion**, and **more interaction**?

The outliers of tomorrow's casino world are those who recognize that **incremental improvements** won't be enough. You must **embrace a visionary mindset**, willing to redefine what a casino can be. This means taking **calculated risks**, investing in **cutting-edge technologies**, and reshaping your casino into a dynamic entertainment ecosystem that offers more than just games—it offers a journey.

So, the question is: *Are you prepared to challenge the status quo? Are you ready to take your casino from being just another player in the market to becoming a true innovator?*

A Visionary Blueprint

To build tomorrow's casino today, you need a **blueprint for the future**—one that looks beyond the current trends and focuses on **long-term innovation**. The casino industry in 2030 and beyond will be defined by how well it integrates **immersive experiences**, **hyper-personalization**, and **real-time engagement**. This blueprint isn't just about adding new games or upgrading facilities—it's about **redefining the entire experience** for your players.

Key Elements

Immersive, Interactive Experiences:
The cornerstone of tomorrow's casino will be its ability to offer **immersive environments** that merge reality with the digital world. Imagine your players walking into a gaming room where the tables, walls, and even the air around them can shift to create entirely new experiences in real-time. **Holograms, mixed reality (MR)**, and **augmented reality (AR)** will bring games to life in ways that are hard to imagine today. Think about the scenes from *The Matrix* where characters slip between real and virtual worlds—this is the kind of immersion your players will expect.

Ask yourself: *How can we start integrating these immersive technologies into our gaming environment now? Can we create spaces that transform dynamically to meet player preferences?*

Hyper-Personalization Powered by AI and Big Data:
Tomorrow's players—like Emma—won't just want a tailored experience; they will demand it. **Artificial intelligence (AI)** will be the driving force behind this hyper-personalization, analyzing vast amounts of player data to predict needs and preferences. Imagine a world where every interaction, every game, every promotion is custom-made for each player in real-time. Your casino will need to be able to anticipate what Emma wants before she even knows it herself.

Have you invested in AI-powered systems that can deliver this level of personalization? Are you leveraging **big data** to truly understand your players, or are you still treating everyone the same?

Seamless Multi-Platform Engagement:
The future of gaming will be **cross-platform**, and casinos will need to ensure that the experience is seamless whether players are at home, on the move, or in-person. Emma

should be able to play games on her mobile app, watch live-streamed tournaments, or place sports bets from anywhere in the world—all while connected to her loyalty program and the broader casino experience.

Is your casino positioned to offer this **multi-platform integration**? Can your players interact with your brand consistently across all channels?

Social and Competitive Gaming:
For younger players, gaming is increasingly social and competitive. **eSports**, **live tournaments**, and **real-time betting** will become key elements of the casino experience. Tomorrow's casinos will host **interactive competitions**, blending live events with virtual gaming worlds where players can compete in real-time, share their victories on social media, and even stream their gaming sessions.

Have you explored how to integrate social and competitive gaming into your casino? Are you ready to build **eSports arenas** and interactive lounges where players can gather and compete?

These are not just elements of a dream—they are the pillars of the blueprint for the **casino of tomorrow**. The challenge is not whether these things will happen but whether you will be one of the casinos that leads the way.

Adapting to a New Generation

The gaming industry is already seeing a **major demographic shift**, which will only accelerate. **Millennials** and **Gen Z** players are redefining what they expect from entertainment, and casinos must be prepared to **adapt** to these changing expectations. These players grew up with video games, social media, and digital experiences—they are looking for something that feels **interactive, social,** and **customized.**

Building for a Socially Connected World

For younger players, gaming is not a solitary activity—it's a **shared experience.** They want to compete with their friends, share their achievements online, and be part of a larger community. Your casino must be able to cater to these desires by offering **social gaming experiences** where players can interact both in-person and virtually.

Are you prepared to offer **live-streaming capabilities** and **social media integration** that allow players to broadcast their gaming experiences? Have you considered making your casino the central hub of a **social gaming network?**

Las Vegas the World's Entertainment Capital

By 2030, Las Vegas will not only be the world's casino capital but also the **global entertainment capital,** offering more than just games. The casinos of the future will need to cater to **sports fans, concert-goers,** and **cultural tourists,** transforming themselves into **multi-dimensional entertainment hubs.**

Integrating Sports and Entertainment

As **sports betting** becomes more widespread and Las Vegas continues attracting major sporting events, your casino must fully integrate sports experiences into its offerings. Imagine having live sports arenas where players can bet on games in real-time, interact with virtual player stats, and experience the game in immersive environments.

It's not just about betting—it's about becoming a destination for **sports entertainment**.

Ask yourself: *How can we make sports betting more interactive, immersive, and engaging? Can we offer a space where gaming and live sports come together seamlessly?*

Offering a Full Entertainment Ecosystem

Future casinos will offer concerts, festivals, and cultural experiences beyond traditional entertainment in addition to sports. Players like Emma will come to your casino not just to gamble but also to attend live performances, interact with virtual celebrity hosts, and engage in **immersive cultural events**.

Have you started building partnerships with **entertainment brands** to offer a full ecosystem of experiences? Can your casino transform into a **destination for all types of entertainment**?

Embracing Sustainability and Responsibility

As casinos expand their offerings and invest in new technologies, they must also lead in **sustainability** and **corporate social responsibility (CSR)**. The future of gaming must be **sustainable**, with casinos adopting eco-friendly building practices, energy-efficient technologies, and responsible gaming policies that protect both the environment and players.

Sustainable Casinos for a New Era

The casino of the future will need to be **eco-conscious**, reducing its carbon footprint through smart building materials, renewable energy sources, and sustainable water management. This isn't just a trend—it's a responsibility that future players will demand.

Is your casino ready to embrace **green design** and sustainability? Are you building an environment that will stand the test of time, both environmentally and ethically?

Responsible Gaming and Data Privacy

With the rise of AI, big data, and hyper-personalization comes a responsibility to protect players from **problematic behaviors**. Tomorrow's casinos must lead in offering **responsible gaming tools**, ensuring that players like Emma can set limits, track their spending, and enjoy a safe, healthy gaming environment. At the same time, player data must be handled carefully, respecting privacy and security.

Are you ready to implement **responsible AI systems** that enhance the player experience and protect their well-being?

The Final Call – Building Tomorrow's Casino Today

The future is here. The **call** to build tomorrow's casino is ringing loud and clear. But not everyone will answer it. The casinos that do will be the ones that thrive in the new era of entertainment, while those that don't risk becoming relics of the past.

Answering the Call

To answer the call and build the casino of tomorrow, you must:

- **Invest in Immersive Technology:**
  Start by investing in the tools and platforms that redefine the gaming experience—holograms, mixed reality, AI, and personalization.
- **Think Beyond Games:**
  Shift your mindset from focusing solely on gaming to offering a complete **entertainment ecosystem**. Build spaces where players can engage in live events, concerts, sports, and cultural experiences.
- **Put Players First with Hyper-Personalization:**
  Use AI and big data to deliver **real-time, hyper-personalized experiences** that anticipate your players' needs and desires, creating a seamless journey from start to finish.
- **Lead in Sustainability and Responsibility:**
  Build an environment that is innovative and **responsible**, focusing on sustainability, data privacy, and ethical gaming practices.
- **Start Building Now:**
  Don't wait until 2030 to start—begin laying the groundwork today. By investing in the future now, you will be positioned to **lead the charge** when the next wave of innovation hits.

Key Takeaways

1. **Building Tomorrow's Casino Today is an Imperative:**
   Innovation, immersive experiences, and cutting-edge technologies will define the casino of the future. The question is, are you prepared to build it?
2. **Immersive Technology is Non-Negotiable:**
   Holograms, mixed reality, AI, and personalization are not optional—they are the building blocks of the future casino experience.
3. **The Next Generation of Players Wants More:**
   Social, competitive, and multi-platform gaming experiences will define the future. Is your casino ready to meet their expectations?
4. **Sustainability and Responsibility Will Define Leadership:**
   The casinos that lead in sustainability and responsible gaming

will set the standard for the entire industry. Will you be one of them?

5. **Answer the Call Now:**
The future of gaming is here. The time to act is now. Will you answer the call and build tomorrow's casino today?

APPENDIX

A Legacy of Innovation
"The Evolution of AI and the Collective Genius of Humanity"

In the vast timeline of human history, certain innovations have acted as catalysts, propelling our species forward in ways that have forever altered the course of civilization. From the mastery of fire to the invention of the wheel, from the development of spoken language to the creation of written communication, each of these breakthroughs has contributed to the tapestry of human progress. They have allowed us to build, to learn, to connect, and to push the boundaries of what is possible.

Today, we stand on the brink of yet another such breakthrough—Generative AI (Gen AI). Much like its predecessors, Gen AI represents the culmination of millennia of human thought, experimentation, and collaboration. It is the latest in a long line of innovations that have fundamentally changed how we interact with the world and with each other. But to fully appreciate the significance of Gen AI, we must first understand the historical context in which it exists. This chapter explores the evolution of artificial intelligence, tracing its roots back to the earliest innovations that have shaped our species and examining how this technology stands as a testament to the collective genius of humanity.

The Dawn of Innovation

The mastery of fire was perhaps the first great leap in human innovation. Fire provided warmth, protection, and a means to cook food, transforming early human societies from vulnerable foragers into capable hunters and gatherers. But fire was more than just a tool for survival—it was the beginning of humanity's journey toward mastery over nature. It allowed our ancestors to expand into new territories, survive in harsher climates, and eventually build the first permanent settlements.

The invention of the wheel followed, marking another pivotal moment in human history. The wheel revolutionized transportation, enabling the movement of goods and people over greater distances than ever before. It facilitated trade, cultural exchange, and the spread of ideas, laying the groundwork for the complex societies that would later emerge. The wheel's impact on human progress cannot be overstated—it was a key enabler of the agricultural and industrial revolutions that would follow millennia later.

But perhaps the most profound innovations in human history were the development of spoken language and written communication. These breakthroughs allowed for the transfer of knowledge across generations, enabling the cumulative learning that has defined our species. Oral language allowed humans to share experiences, teach each other, and coordinate complex activities, while written language preserved knowledge for future generations. Without these tools, the rapid pace of change we see today would have been impossible.

Catalyst for Knowledge Transfer

The ability to record and transmit knowledge across time and space was a game-changer for humanity. With the advent of written language, civilizations could document their discoveries, laws, and stories, ensuring that each generation could build upon the achievements of those who came before. The Sumerians, who invented cuneiform writing around 3200 BCE, were among the first to harness this power. Their ability to record transactions, legal codes, and astronomical observations marked the beginning of recorded history.

Written language also enabled the creation of complex societies, where laws and governance could be codified and enforced. The ancient Egyptians, Greeks, and Romans all relied on written records to administer their vast empires, from tax collection to military logistics. The preservation of knowledge through written texts allowed these civilizations to flourish, laying the intellectual foundations for the Renaissance, the Enlightenment, and eventually, the modern world.

The Gutenberg printing press, invented in the mid-15th century, represents another quantum leap in the dissemination of knowledge. By making books more accessible and affordable, the printing press democratized information, fueling the spread of ideas across Europe and igniting the flames of the Reformation and the Scientific Revolution. The rapid exchange of knowledge that followed set the stage for the Industrial Revolution, which would transform society in ways that were previously unimaginable.

From Mechanical Minds to Machine Learning

The concept of artificial intelligence—machines that could mimic human thought—has been a part of human imagination for centuries, long before it became a scientific pursuit. Early myths and legends often featured automata, mechanical beings created to perform tasks or even think like humans. These stories reflect a deep-seated fascination with the idea of creating life and intelligence through technology.

The first concrete steps toward artificial intelligence came in the mid-20th century, during the height of the digital revolution. The development of computers, which could perform calculations far beyond the capacity of the human brain, laid the groundwork for the field of AI. In 1950,

British mathematician Alan Turing published his seminal paper, "Computing Machinery and Intelligence," in which he proposed the idea of machines that could "think" and introduced the famous Turing Test as a measure of machine intelligence.

The 1950s and 1960s saw the emergence of the first AI programs, which were primarily rule-based systems that could solve specific problems, such as playing chess or proving mathematical theorems. These early AI systems were impressive but limited in scope, as they relied on pre-programmed rules and lacked the ability to learn from experience.

The next major leap in AI came with the development of machine learning in the late 20th century. Unlike earlier AI systems, which required explicit programming for each task, machine learning algorithms could analyze large datasets, identify patterns, and make predictions or decisions based on that data. This shift marked a turning point in the field, as it allowed AI systems to improve over time and tackle more complex, real-world problems.

The rise of machine learning led to breakthroughs in fields such as natural language processing, computer vision, and robotics. These technologies began to permeate various aspects of daily life, from voice-activated virtual assistants to facial recognition software. Yet, even as AI became more powerful, it remained largely a tool for solving specific, narrowly defined tasks.

A New Paradigm

Generative AI, the latest development in the field of artificial intelligence, represents a significant shift from earlier approaches. While traditional AI systems were designed to analyze data and make decisions, generative AI models can create entirely new content—whether text, images, music or even complex simulations. These models, such as OpenAI's GPT-3, can generate human-like language and produce creative works that were once thought to be the exclusive domain of human intelligence.

Generative AI models are built on deep learning algorithms that process vast amounts of data, learning the nuances of language, art, and other forms of expression. They can generate coherent and contextually relevant content, complete complex tasks such as writing essays, composing music, or designing graphics, and even engage in natural and human-like conversations.

The implications of generative AI are profound. For the first time in history, machines are not just processing information but creating it. This has the potential to revolutionize industries ranging from entertainment and design to education and communication. Generative AI can automate the creation of content, assist in brainstorming and ideation, and provide personalized experiences at an unprecedented scale.

But generative AI is not just a technological breakthrough—it is the culmination of thousands of years of human innovation. It is built on the foundations laid by the discovery of fire, the invention of the wheel, the development of spoken and written language, and the accumulation of knowledge across generations. Each of these innovations expanded the horizons of what humans could achieve, and generative AI is no different. It represents the next step in our journey toward mastering the tools of creation and harnessing the collective intelligence of humanity.

Gen AI as a Collective Effort of Human Knowledge

To understand the significance of generative AI, we must recognize it as the result of a collective effort that spans millennia. The development of AI and machine learning is not just the work of a few brilliant minds—it is the culmination of countless contributions from scientists, engineers, mathematicians, and thinkers throughout history.

The knowledge that powers generative AI models has been accumulated over thousands of years and passed down through written texts, scientific discoveries, and technological advancements. From the early mathematicians who laid the foundations of logic and computation to the pioneers of computer science in the 20th century, each generation has built upon the achievements of the previous one, contributing to the development of AI as we know it today.

Generative AI is a reflection of this collective intelligence. It leverages the vast store of human knowledge encoded in digital form—books, articles, images, and more—to generate new content and insights. It is, in a sense, the embodiment of the knowledge and creativity of our species, distilled into algorithms that can learn, adapt, and create.

But generative AI also represents a new chapter in the history of human innovation. It is a tool that can amplify our abilities, allowing us to create more, learn more, and connect more deeply with each other. Just as the invention of written language enabled the transfer of knowledge across generations, generative AI has the potential to accelerate the pace of innovation, opening up new possibilities for what we can achieve as a species.

As we enter the era of generative AI, it's more than just a leap in technology—it's the next chapter in humanity's story, one that began with fire, the wheel, and the birth of language. GenAI stands at the crossroads of these innovations, holding the potential to reshape creativity, knowledge, and connection in ways we've only begun to imagine. Our next book will explore how this powerful tool can redefine the future.

Relevant AI Models

The evolution of AI frameworks and algorithms over the past two decades has set the stage for practical applications in the casino industry. Like many others, the casino industry is rapidly evolving, and data-driven insights have become essential to maintaining a competitive edge. Today's casinos are not just spaces for gambling—they are entertainment hubs that blend gaming with leisure, fine dining, and live entertainment. However, this diverse offering requires understanding players on a deeper level than ever before. A data-driven approach enables casinos to engage players meaningfully, enhance their experiences, and optimize marketing efforts.

AI models help casinos interpret massive amounts of data, revealing patterns and insights that would be impossible to see through human analysis alone. They can categorize players, forecast behaviors, personalize offers, and refine marketing strategies. While several types of AI models are available, choosing the right ones—and knowing when to use each—is critical to getting optimal results. This chapter explores eight AI models that can significantly impact a casino's marketing and engagement strategy, providing a blueprint for integrating each into a comprehensive, data-driven approach to player engagement.

Each model discussed has its strengths and potential limitations. While some excel at segmentation, others are valuable for real-time recommendations or behavioral predictions. Together, these AI models form an interconnected system that, when implemented strategically, can enhance player satisfaction, build loyalty, and ultimately drive revenue growth. Here is an in-depth look at each AI model, detailing its purpose, applications in casino marketing, and how it can transform player engagement.

Here's how a few foundational and generative models stack up (check **Visit bit.ly/player360** for more details:

Model	Purpose	Application	Pros	Cons
Clustering (K-Means, GMM)	Group players by behavior/preferences	Player segmentation (e.g., VIPs, casuals)	Accurate, creates valuable personas	Needs large dataset
Collaborative Filtering	Recommend based on similar players	Personalized game and suggestions	Accurate for existing players, scalable	Limited for new players
Content-Based Filtering	Suggest games with similar attributes	Engage new players with familiar game types	Effective for new players, easy to interpret	Limited variety, needs structured data
Hybrid Models	Combine collaborative and content-based data	Comprehensive personalization	High accuracy, works for all player types	Complex setup, high computational cost
Predictive Models (XGBoost, Logistic)	Forecast player response/campaign success	Target campaigns with high response likelihood	Clear ROI, interpretable results	Requires historical data, frequent updates
Deep Learning (RNNs, Transformers)	Real-time personalization	Dynamic, responsive offers and engagement	Captures complex patterns, real-time	Expensive, data-intensive
NLP Models (BERT, LDA)	Analyze player feedback/sentiment	Sentiment-driven campaign improvements	Real-time insights, trend identification	Complex, needs large text datasets
Multi-Armed Bandits	Test and refine campaign effectiveness	Optimize campaign strategies dynamically	Clear insights, flexible adj	Resource-heavy, can limit reach

References

Here is a list of leading journals and publications that cover various aspects of AI, technology, and their applications in industries such as alcohol and beverage. These resources are valuable for staying up to date on the latest research, trends, and insights in the field:

Journal of Artificial Intelligence Research (JAIR)

Focus: This peer-reviewed journal covers various AI topics, including machine learning, natural language processing, robotics, and AI applications. It's a key resource for understanding the latest academic research and developments in AI.

Website: www.jair.org

IEEE Transactions on Neural Networks and Learning Systems

Focus: Published by the IEEE, this journal focuses on the theory, design, and application of neural networks and learning systems. It's particularly relevant for those interested in the technical and algorithmic aspects of AI.

Website: ieeexplore.ieee.org

Artificial Intelligence Journal

Focus: This is one of the oldest and most respected journals in the AI field, covering both theoretical and applied aspects of AI. Topics include AI methodologies, systems, and emerging technologies.

Website: www.journals.elsevier.com/artificial-intelligence

Machine Learning Journal

Focus: This journal publishes high-quality research on all aspects of machine learning, including the development and application of learning algorithms. It's a critical resource for understanding the mechanisms behind AI systems.

Website: www.springer.com/journal/10994

AI & Society

Focus: This interdisciplinary journal explores the social implications of AI, including ethical, cultural, and philosophical issues. It's an excellent source for understanding the broader impact of AI on society and industry.

Website: www.springer.com/journal/146

MIT Technology Review

Focus: Although not a peer-reviewed journal, MIT Technology Review is a leading publication that covers the latest trends and developments in technology, including AI. It's accessible and provides insights into how AI is shaping various industries, including alcohol and beverage.

Website: www.technologyreview.com

Nature Machine Intelligence

Focus: A prestigious journal from the Nature Publishing Group, this publication focuses on AI, machine learning, and robotics, covering both fundamental research and applications. It's a go-to resource for cutting-edge AI research.

Website: www.nature.com/natmachintell

The AI Magazine

Focus: Published by the Association for the Advancement of Artificial Intelligence (AAAI), this magazine provides broad coverage of AI research, applications, and the field's history and impact. It's particularly useful for those looking to stay informed about AI developments in a more digestible format.

Website: www.aaai.org/Magazine/magazine.php

Harvard Business Review (HBR)

Focus: HBR regularly publishes articles on AI's impact on business, strategy, and management. While not exclusively focused on AI, it's invaluable for understanding how AI intersects with business leadership and operational strategy.

Website: hbr.org

International Journal of Robotics Research (IJRR)

Focus: While focused on robotics, this journal often covers areas of AI that overlap with robotics and autonomous systems, making it relevant for those interested in the AI-driven automation aspects of the industry.

Website: journals.sagepub.com/home/ijr

Journal of Business Research

Focus: This journal occasionally publishes research on AI applications in business contexts, including customer engagement, marketing, and operations. It's a valuable resource for understanding the business implications of AI.

Website: www.journals.elsevier.com/journal-of-business-research

Journal of Data Science

Focus: This journal covers topics related to data science, including the role of AI in data analysis, machine learning, and the implications for industries such as alcohol and beverage. It's a resource for those looking to understand how data drives AI applications.

Website: jds-online.org

These journals and publications are essential for keeping abreast of the latest research, trends, and insights in AI and its application across various industries. By regularly consulting these sources, you can stay informed and ensure that your AI initiatives are grounded in the most current and relevant knowledge.

Book Recommendations

1. The Experience Economy" by B. Joseph Pine II and James H. Gilmore

2. "Predictive Analytics: The Power to Predict Who Will Click, Buy, Lie, or Die" by Eric Siegel

3. "The Loyalty Leap: Turning Customer Information into Customer Intimacy" by Bryan Pearson

4. "Hooked: How to Build Habit-Forming Products" by Nir Eyal

5. "The Tipping Point: How Little Things Can Make a Big Difference" by Malcolm Gladwell

6. "The Digital Transformation Playbook: Rethink Your Business for the Digital Age" by David L. Rogers

7. "Data-Driven: Creating a Data Culture" by Hilary Mason and DJ Patil

8. "Competing in the Age of AI: Strategy and Leadership When Algorithms and Networks Run the World" by Marco Iansiti and Karim R. Lakhani

9. "Bold: How to Go Big, Create Wealth, and Impact the World" by Peter H. Diamandis and Steven Kotler

10. "The Infinite Game" by Simon Sinek

Acknowledgments

Writing Player 360 has been like playing the most exhilarating and challenging football game of my life—with an incredible team, supportive coaches, and lessons learned on and off the field. This book would not exist without the collective effort of those who played vital roles.

First, my co-author, Bryan—your strategic mind, dedication, and unwavering support made you the perfect quarterback on this journey. To my mentors and friends who offered invaluable support and wisdom, Jacob Lanning, your sharp analysis pushed us to refine our approach. Mark Wayman, you revealed the vast potential of the gaming world. Cliff Atkins, Charly Paelink, Ramesh Srinivasan your early support and friendship set me on this path.

Richard Marcus, your encouragement to share our insights and detailed feedback turned an ambitious project into reality. Gavin Isaac and Stephen Moore, your guidance inspired new perspectives. John Wolfe, your lessons on life left a mark on me. Rahul and Thomas, you have shaped my leadership. Amit, you were the quarterback leading with resilience. Yogi, your foundational support was vital to every move forward. To the unsung heroes who supported us throughout this journey, thank you. Your belief and contributions fueled this project.

To my readers—you are the reason behind this work. Your time, thoughts, and feedback mean the world to me.

With deep gratitude,
Sanjiv

ELEVATE PLAYER ENGAGEMENT

Transform Your Casino Strategy with Player 360 Resources

To help you take full advantage of the insights in *Player 360*, we've developed two powerful resources that you can download for free:

Player 360 Self-Assessment
Evaluate your current strategies and pinpoint areas for growth.
Use this comprehensive self-assessment to measure how well your current approach aligns with industry best practices.

Player 360 Blueprint
Turn strategy into action with this step-by-step guide.
This Blueprint walks you through each phase of the Player 360 framework, offering detailed steps for implementing data-driven, personalized, and multi-channel player experiences.

To download these resources, simply:

1. **Scan the QR Code Below**
or
2. **Visit bit.ly/player360**

Note: You'll be prompted to enter your email to receive the email with the download link.